DEPARTMENT OF THE INTERIOR
HUBERT WORK, Secretary

UNITED STATES GEOLOGICAL SURVEY
GEORGE OTIS SMITH, Director

Bulletin 718

GEOLOGY AND ORE DEPOSITS OF THE CREEDE DISTRICT, COLORADO

BY

WILLIAM H. EMMONS.

AND

ESPER S. LARSEN

WASHINGTON
GOVERNMENT PRINTING OFFICE
1923

CONTENTS.

ILLUSTRATIONS.

INSERT.

GEOLOGY AND ORE DEPOSITS OF THE CREEDE DISTRICT, COLORADO.

By W. H. EMMONS and E. S. LARSEN.

CHAPTER I.—INTRODUCTION.

LOCATION AND TOPOGRAPHY.

The Creede mining district is in Mineral County, southwestern Colorado, near the eastern border of the elevated region that is generally known as the San Juan Mountains. The town of Creede is on Willow Creek a few miles above its junction with the Rio Grande.

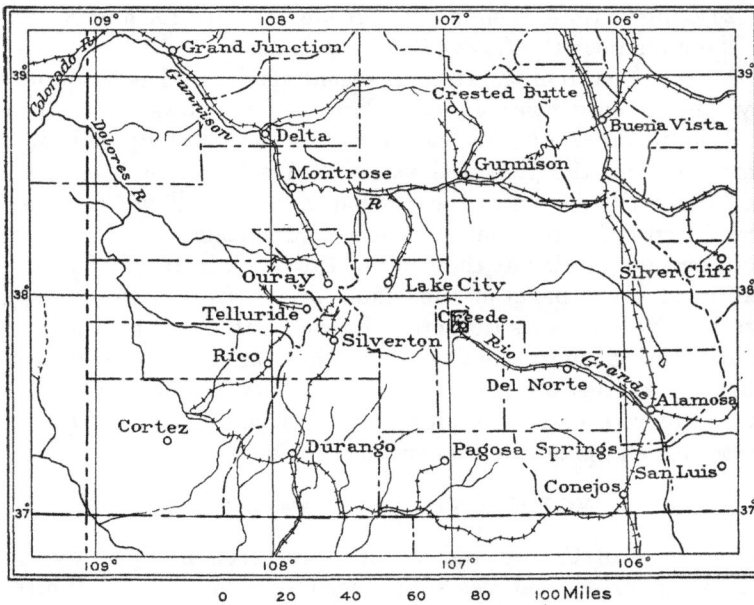

FIGURE 1.—Index map of a portion of Colorado showing location of Creede mining district.

The area shown on the accompanying maps of Creede and vicinity measures about 4½ miles from east to west and 5¾ miles from north to south. Its position with respect to other mineralized areas in Colorado is shown on figure 1. The lowest part of this area, about a mile downstream from Creede, is about 8,700 feet above sea level, and the highest point is the summit of Nelson Mountain, near the

1

north border of the area, which is over 12,050 feet above sea level. The relief is therefore over 3,300 feet. The points mentioned are shown on Plates I and II (in pocket).

Only a few miles to the north is San Luis Peak, which is more than 14,000 feet above the sea.

The area may be divided into four topographic units—the valley of the Rio Grande, with its gently rolling grass-covered hills; the sharp, rugged canyons of the main streams; the gentle slopes in the upper parts of the drainage basins; and the rounded slopes and mesas of the uplands. The valley of the Rio Grande, which is cut in the soft Creede formation, is about 2 miles wide near Creede, which is on its north border. It is shown on Plate III, with the rolling hills of tuff and the alluvial flat of the stream.

The canyons, especially the canyon of Willow Creek above Creede, present sheer cliffs several hundred feet high, nearly vertical and locally overhanging. Near the Monte Carlo mine for 1,000 feet vertically the cliffs have an average slope of about 60°. Plate VIII, B (p. 24), shows these cliffs above Creede. Plates IX and X (p. 24) show the cliffs near the forks of Willow Creek.

In the upper parts of the streams recent glaciation has modified the valleys and given gentler slopes. Plate IV, A, shows the canyon of West Willow Creek on the left of the picture, the uplands in the background, and the glaciated valley of West Willow Creek on its right. Plate IV, B, shows the canyon of East Willow Creek in the central foreground, the mesa of Nelson Mountain in the left background, and mountains at the head of the creek in the distance.

Like most other parts of the mountainous area of the San Juan region, the Creede area is timbered and well watered. It is served by a broad-gage branch of the Denver & Rio Grande Western Railroad, which daily carries sleeping cars from Denver and Pueblo. The district is more accessible than many other camps of the San Juan, and rates for shipping ore are considerably lower.

FIELD WORK AND ACKNOWLEDGMENTS.

The geology of Creede and vicinity has been mapped by E. S. Larsen, who for several years has been associated with Whitman Cross in the study of the general geology of the San Juan Mountains. Mr. Larsen spent several brief periods in the study of the geology of the region near Creede incidental to work in the San Cristobal quadrangle, the border of which lies a short distance west of Creede, but the larger part of his work was done during the summer of 1911. W. H. Emmons was detailed to study the underground workings and the ore deposits, and spent about seven weeks, from July 28 to September 15, 1911, in this work. In 1912 also he spent

CREEDE, COLO.

A. BASIN OF WEST WILLOW CREEK.

B. BASIN OF EAST WILLOW CREEK

four weeks in the field. A preliminary report [1] on the geology and ore deposits of the district appeared in 1912.

The thanks of the writers are due to the mining operators of the district for friendly cooperation and support, especially to the officers of the Solomon mines and to Messrs. S. B. and Albert Collins for the use of mine maps and to Messrs. William Barnett, G. B. Gleason, and J. F. Wilson for many favors.

HISTORY.

In the eighties the upper part of the valley of the Rio Grande was a route of transportation between Wagonwheel Gap and the flourishing camps near Silverton and Lake City. This route passed very near the present site of Creede and nearer still to Sunnyside, a small camp about 2 miles west of Creede. Some of the prospectors halted at Sunnyside to investigate the steep mountain slopes along the valley and, finding encouraging indications, located several claims. J. C. MacKenzie and H. M. Bennett located the Alpha claim at Sunnyside April 24, 1883, and with James A. Wilson located the Bachelor claim, near the present site of Creede, July 1, 1884. Some prospecting was done in the middle eighties, principally at Sunnyside, and futile attempts were made to work the ores in arrastres. As early as 1885 Charles F. Nelson prospected the site of Creede and, getting small returns, went to Sunnyside. Richard and J. N. H. Irwin bought the Alpha claim in 1885 and located other claims near by. There is no record of any new discoveries from 1886 until August, 1889, when N. C. Creede, E. R. Naylor, and G. L. Smith located the Holy Moses claim on Campbell Mountain. The next summer Creede located the Ethel and C. F. Nelson located the Solomon claim. The mining district that was formed was called the King Solomon district; it is east of and nearly continuous with the Sunnyside district.

The promising assays that were obtained from the Holy Moses claim, which was christened from the exclamation of Creede when he first beheld the outcrop, led to the rapid development of the district. When it became generally known that Creede had sold an interest in the Holy Moses mine to D. H. Moffat and associates, of Denver, prospecting was renewed with great vigor.

In June, 1891, Theodore Renniger and Julius Haas, grubstaked by Ralph Granger and Eric Buddenbock, two butchers of Del Norte, set out to prospect the region of Creede. It is said that a search for their strayed burros led Renniger to the outcrop on the Last Chance claim, which was located in August, 1891. Creede, who was then engaged in developing his claims on East Willow Creek, visited the

---

[1] Emmons, W. H., and Larsen, E. S., A preliminary report on the geology and ore deposits of Creede, Colo.: U. S. Geol. Survey Bull. 530, pp. 1-26, 1912.

site of the discovery and traced the outcrop for some distance. Impressed with the surface indications Creede prevailed on Renniger to define his claim (the Last Chance), and then located next to it the Amethyst claim, in the names of D. H. Moffat, L. E. Campbell, and himself. Haas sold his interest in the Last Chance claim to his partners for $10,000, and in November, 1891, Renniger and Buddenbock sold their thirds for $65,000 to investors in Leadville and Denver. Granger, one of the original locators, was offered $100,000 for his third interest but did not sell.

In April, 1891, J. C. MacKenzie and W. V. McGilliard located the Commodore, and in August of that year George K. Smith and S. D. Coffin located the New York as the southerly extension of the Last Chance. This was on the Amethyst or "Big" vein, upon which the Bachelor claim had been located six years before, but the two locations were nearly three-quarters of a mile apart. Within a few months the Amethyst vein was pegged for a distance of nearly 2 miles along its strike. Mammoth, Campbell, and MacKenzie mountains each received a due share of attention from numerous prospectors.

The finding of rich ore almost contemporaneously in the discovery shafts of the Last Chance, Amethyst, and New York was widely heralded. Late in 1891 and in 1892 Creede experienced a boom that rivals anything in the earlier history of western mining camps. For a mile and a half the narrow gulch of Willow Creek was the scene of active building, and within a few months a city of perhaps 10,000 people had grown up. The Denver & Rio Grande Railroad, which had its terminus at Wagonwheel Gap, was extended to Creede, and the first train arrived December 16, 1891. The principal part of the town, which was at first called Jimtown, was north of Wall Street. Owing to the small area of flat land available this part was built up almost solidly.

The land along Willow Creek south of Wall Street was a school section leased by the State of Colorado to Maj. M. V. B. Wason. The State, having obtained a release from Maj. Wason, sold the land at auction for town lots. The auction was conducted February 27, 1892, by Governor Routt, who was president of the State land board. Lots on Main Street sold as high as $2,500 each. It is said that over $50,000 cash was paid for land at the sale, and this represented only 30 per cent of the purchase price. This addition was known as South Creede. It was later added to Jimtown, and the two were incorporated as a city. Only about three months after the lot sale a disastrous fire burned practically all houses north of Wall Street, which had been the principal section of the city.

By December 31, 1892, the district had produced 46,365 tons of ore, valued at $4,215,000.[2] Of this the Amethyst mine produced nearly one-half and the Last Chance more than one-third.

---

[2] Mines and mining of Colorado, a souvenir volume published at Denver.

The district has been producing almost continuously since the advent of the railroad, and in the nineties the daily output was large. During some of these years the price of silver was very low, but nevertheless the mining operations were profitable. Successful operations were in progress even during the lean years of 1893 and 1894. As shown by the table on page 10, the total production of Mineral County is valued at $41,698,374. Practically all of this is from the Creede district. This includes, in order of their value, silver, lead, gold, and zinc. About half this sum was paid as dividends, notwithstanding the low prices at which the metals were marketed. The Creede district ranks with Tonopah, Nev., as one of the most productive silver camps in the United States developed after the great slump in the price of silver. The mines on the Amethyst vein supplied over 90 per cent of the total production and paid an even larger proportion of the dividends.

As shafts were sunk deeper on the Amethyst, Last Chance, and Commodore claims, considerable flows of water were encountered, and owing to the high cost of pumping the need for a deep drainage adit became urgent. Early in the nineties Charles F. Nelson organized the Nelson Tunnel Co. and drove an adit toward the vein, entering Bachelor Mountain just south of the Bachelor claim. This adit was intended to develop some veins that crop out high on Bachelor Mountain. It was driven northwestward about 2,100 feet but did not encounter any workable ore.

Later the Wooster Tunnel Co. was organized and secured a right of way through the Nelson tunnel. This company drove northward from the Nelson tunnel to the Amethyst mine. Contracts were made with the Last Chance, New York, and Amethyst companies to haul their ores, and in return for the advantage of a deep drainage adit some of the companies agreed to pay royalties on ore and waste hoisted through their own shafts. The Commodore mine, which occupied a position topographically more favorable, did not enter into contract with the Wooster Tunnel Co., but drove its own deep adit about 2,200 feet to the vein at a level about 45 feet above the Wooster tunnel. Still later the Wooster tunnel was extended northward, as the Humphreys tunnel, along the Amethyst vein from the Amethyst mine to the Happy Thought and Park Regent mines. This adit, the Nelson-Wooster-Humphreys tunnel, is now about 11,100 feet long, and 9,000 feet of it is driven on or near the Amethyst vein. Its track extends as a surface tramway about 2,400 feet beyond the portal to the Humphreys mill. It is equipped with mule haulage and is the main artery of transportation for all the mines that operate on the Amethyst lode except the Commodore.

MINING AND TREATMENT OF ORES.

General conditions.—Conditions are favorable for cheap mining, as the veins are nearly everywhere of good width. They have been subjected to very extensive fracturing and crushing, so that much of the work has been done with pick and shovel. Much ore in one of the largest stopes was run out or "milled" from the bottom without previous blasting or breaking. At present nearly all the stoping is done with hand drills. Owing to the fractured condition of the rock the miners find their labor lighter than in many neighboring districts.

The mines on the Amethyst and Solomon veins are served by adits. The topography is so rugged that these gain depths of 1,000 to 1,400 feet within comparatively short distances. The mines supply a large quantity of water, which is drained through the adits and is used for milling. Although several deep shafts have been sunk, at present nearly all the ore is passed through the adits.

The larger part of the ore at Creede was partly or completely oxidized. Such ore is not suitable for mechanical concentration and was shipped without dressing to smelters. Most of it has gone to the plant of the American Smelting & Refining Co. at Pueblo. Owing to its siliceous character, favorable rates are obtained for smelting.

In the lower levels of the mines on the north end of the Amethyst lode, especially in the Amethyst and Happy Thought mines, large bodies of sulphides suitable for concentration were encountered. This material was dressed in the Humphreys mill, at North Creede, and in the Amethyst mill, near the Amethyst mine. The Humphreys and Amethyst mills are of the same general type, the equipment including crushers, rolls, classifiers, jigs, tables, and canvas plants. The zinc blende and galena are readily separated, giving clean concentrates, and the pyrite is not so abundant as to reduce the grade of zinc concentrates greatly. Gold and silver are recovered mainly with the lead concentrates or in the slimes. Two smaller mills, the Solomon and the Ridge, have been erected on East Willow Creek about 1½ miles above North Creede. The ore of the Solomon vein is very soft, and some of it is crushed by rolls without preliminary breaking in a jaw crusher.

Humphreys mill.—The Humphreys mill, at North Creede, is the largest in the district. It was built in 1901 pincipally to treat the ore of the Happy Thought mine. Working at full capacity it will treat about 250 tons in 24 hours. The flow sheet of the mill is shown as figure 2. The mill is equipped with water power, but steam is used in winter when the flow of water is insufficient.

The air supplied to the mines is carried 2 or 3 miles through a 4-inch pipe and furnishes power also for the Amethyst mine.

Amethyst mill.—The Amethyst mill is on West Willow Creek, a short distance from the Amethyst mine, with which it is connected by an aerial gravity tramway. The mill is smaller than the Humphreys mill, but the method of treatment employed is closely similar.

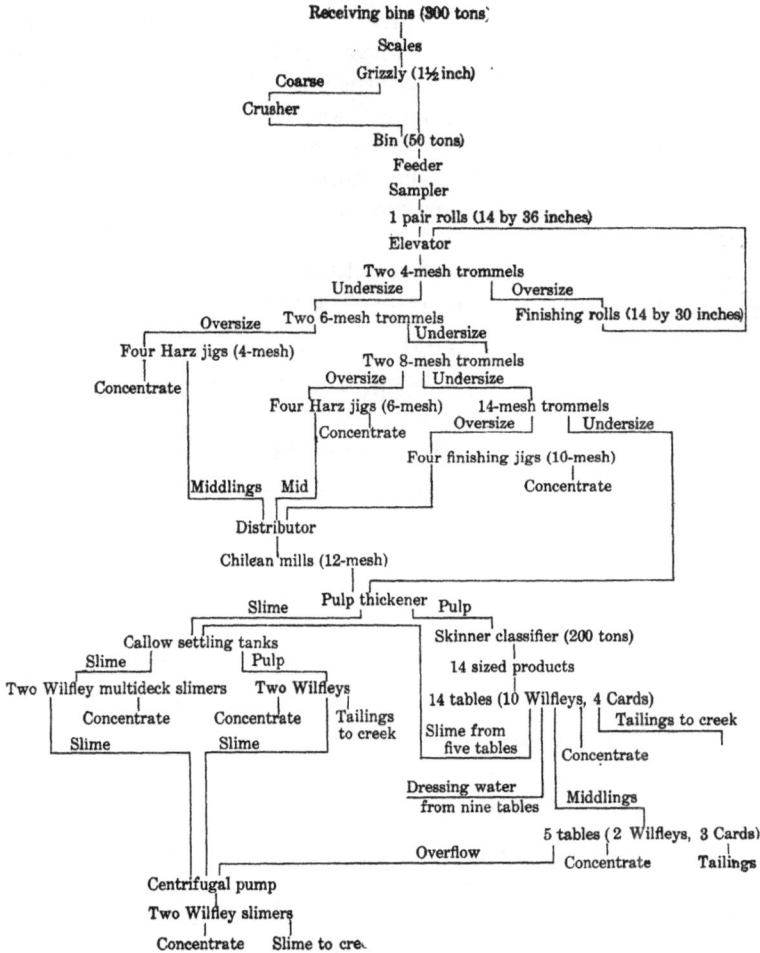

FIGURE 2.—Flow sheet of Humphreys mill.

Solomon mill.—The Solomon mill is on East Willow Creek near the portal of the Solomon adit. Its equipment includes crusher, rolls, screens, jigs, tables, and canvas plant. The general plan of concentration is similar to that employed in the Humphreys and Amethyst mills. Power is supplied by steam. The ore treated carries galena, sphalerite, cerusite, and anglesite. Much of the ore is stiff green chloritic clay, requiring very little grinding. A fair saving is made in jigs and on tables. The slimes that are separated by means of the canvas plant carry lead sulphide and some gold. The Solomon ores contain relatively little silver.

Ridge mill.—The Ridge mill is on East Willow Creek at the portal of the Ridge tunnel, a short distance above the Solomon mill. As in the Solomon mine, the ore is mainly lead and zinc sulphides in a stiff green chloritic clay or gouge. Silver and gold are but sparingly present. Not much of the ore is crushed. The large boulders as a rule carry very little lead and zinc and are thrown away. Some large masses of galena and sphalerite are sorted out by hand. The soft chloritic material carrying disseminated galena and sphalerite is fed to a revolving disintegrating screen. From this it passes through trommel screens to jigs and tables. The mill has a capacity of about 50 tons in 24 hours and is provided with water power which operates a 4-foot Pelton wheel at a pressure of 35 pounds to the square inch.

PRODUCTION.

The subjoined table is taken from a forthcoming work by Charles W. Henderson on the history of mining in Colorado, to be published as a professional paper of the United States Geological Survey. Practically all the metals produced in Mineral County come from the Creede district.

Gold, silver, copper, lead, and zinc produced in Mineral County, Colo., 1889–1920.

Year	Ore (short tons)	Lode gold	Silver — Quantity (fine ounces)	Silver — Avg. price per ounce	Silver — Value	Copper — Quantity (pounds)	Copper — Avg. price per pound	Copper — Value	Lead — Quantity (pounds)	Lead — Avg. price per pound	Lead — Value	Zinc — Quantity (pounds)	Zinc — Avg. price per pound	Zinc — Value	Total value
1889 a	……	……	……	$0.94	……	……	$0.135	……	……	$0.039	……	……	$0.05	……	……
1890 b	……	……	……	1.05	……	……	.156	……	……	.045	……	……	.055	……	……
1891	……	c d $10,055	c d 378,899	.99	$374,382	……	.128	……	c d 354,854	.043	$15,259	……	.05	……	$399,695
1892	……	c e 87,219	c 2,391,514	.87	2,080,617	……	.116	……	c e 3,000,000	.04	120,000	……	.046	……	2,287,836
1893	……	e 53,252	c 4,897,684	.78	3,820,194	……	.108	……	c 7,500,000	.037	277,500	……	.04	……	4,150,946
1894	……	c 40,336	c 1,866,927	.63	1,176,164	……	.095	……	c f 6,500,000	.033	214,500	……	.035	……	1,431,000
1895	……	e 114,482	c 1,423,038	.65	924,975	……	.107	……	c f 8,220,870	.032	263,068	……	.036	……	1,302,525
1896	……	c 52,238	c 1,560,865	.68	1,061,388	……	.108	……	c f 6,021,109	.03	180,633	……	.039	……	1,294,259
1897	……	h 61,328	h 3,070,576	.60	1,842,346	h 1,500	.12	$180	h 6,080,673	.036	218,904	……	.041	……	2,122,758
1898	……	h 46,383	h 4,177,184	.59	2,464,539	h 14,729	.124	1,826	h 5,453,104	.038	207,218	f i 200,000	.046	$9,200	2,729,166
1899	……	h 91,671	h 3,796,899	.60	2,278,139	h 20,223	.171	3,458	h 5,677,162	.045	255,472	f i k 100,000	.058	5,800	2,634,540
1900	……	h 209,387	h 2,280,038	.62	1,413,623	h 2,614	.166	434	h 14,951,956	.044	657,886	f i l 450,000	.044	19,800	2,301,130
1901	……	h 102,813	h 1,816,023	.60	1,089,614	h 1,007	.167	168	h 10,519,895	.043	452,355	f i l 1,800,000	.041	73,800	1,718,750
1902	……	h 112,838	h 1,923,973	.53	1,019,706	……	.122	……	h 9,291,358	.041	380,946	h 2,047,555	.048	98,283	1,611,773
1903	……	h 178,961	h 1,608,788	.54	866,746	h 133	.137	18	h 8,600,646	.042	361,227	h 2,634,000	.054	142,236	1,551,188
1904	m 124,278	m 222,864	h 1,664,633	.58	965,487	h 1,337	.128	171	h 13,346,436	.043	573,897	h 4,402,697	.051	224,538	1,986,957
1905	m 91,338	m 216,994	h 1,198,442	.61	728,000	h 107	.156	17	h 11,889,797	.047	558,397	h 2,515,628	.059	148,422	1,651,830
1906	m 126,164	m 176,150	h 1,254,058	.68	852,759	……	.193	……	h 14,886,356	.057	848,522	h 2,892,061	.061	176,416	2,053,847
1907	m 104,977	m 142,803	h 1,246,961	.66	822,994	h 12,711	.20	2,542	h 12,980,288	.053	687,955	h 2,691,216	.059	158,782	1,815,076
1908	m 61,131	m 127,549	h 830,951	.53	440,404	h 41	.132	5	h 8,235,025	.042	345,997	m 1,100,107	.047	51,705	965,660

a Emmons, W. H., and Larsen, E. S., A preliminary report on the geology and ore deposits of Creede, Colo.: U. S. Geol. Survey Bull. 530, p. 43, 1912. Holy Moses mine located August, 1889.

b In June, 1891, Last Chance deposit discovered; later the Amethyst. First train arrived at Creede, December 16, 1891.

c From reports of the agents of the Mint in annual reports of the Director of the Mint, the gold and silver are prorated to correspond with the corrected figures of the total production of the State by the Director of the Mint. The lead is prorated to correspond with the total production of lead for the State as given in annual volumes of Mineral Resources, any "unknown production" of the State being distributed proportionately to the several counties. The copper is treated similarly to the lead, but as Mineral Resources figures for copper include copper from matte and ores treated in Colorado smelters from other States, the copper figures are subject to revision.

d For 1891, in Mint report, Mineral County output is found (p. 184) under Saguache County. Separation is made as closely as possible from data available.

e For 1892, in Mint report, Mineral County output is found under Hinsdale County (p. 126) and under Rio Grande County (p. 129). Director of the Mint gives for production of Ethel and Holy Moses mines $65,220 in gold and 59,317 ounces of silver: for Amethyst, Bachelor, Del Norte, and Last Chance $25,932 in gold and 2,366,778 ounces of silver. S. F. Emmons (U. S. Geol. Survey Mineral Resources, 1892, p. 68, 1893) says: "Creede credited by Mint authorities with $3,500,000 (at coining rate of $1.29+) of silver in 1892, other estimates giving even a larger amount." Mineral Industry, vol. 1, p. 177, 1892, says: "Output estimated at about 5,000,000 ounces. Last Chance and Amethyst mines largest producers."

f Estimated by C. W. Henderson.

g Mint gives 1,412,226 pounds of lead.

h Colorado State Bureau of Mines, being figures of smelter and Mint receipts.

i Mineral Industry, 1896, p. 727: "A small amount of zinc blende concentrates was shipped in 1898 from a mine at Creede."

j Estimated by C. W. Henderson to correspond with total production of State.

k Mineral Industry, 1899, p. 636: "Considerable (zinc) ore was shipped from Creede, Leadville, and other points to the zinc smelters for experiments."

l Mineral Industry, 1901, p. 651, on basis of metallic content of the ore produced, gives 2,088,000 pounds.

m U. S. Geol. Survey Mineral Resources.

Gold, silver, copper, lead, and zinc produced in Mineral County, Colo., 1889-1920—Continued.

Year.	Ore (short tons).	Lode gold.	Silver.			Copper.			Lead.			Zinc.			Total value.
			Quantity (fine ounces).	Average price per ounce.	Value.	Quantity (pounds).	Average price per pound.	Value.	Quantity (pounds).	Average price per pound.	Value.	Quantity (pounds).	Average price per pound.	Value.	
1909	m 64,941	m $108,825	m 891,185	$0.52	$463,416	m 17,401	$0.13	$2,262	m 9,096,816	$0.043	$388,583	m 1,817,296	$0.054	$98,134	$1,061,220
1910	m 62,956	m 121,181	m 773,722	.54	417,810	m 29,031	.127	3,687	m 8,246,000	.044	362,824	m 2,421,926	.054	130,784	1,036,286
1911	m 65,932	m 179,196	m 545,319	.53	289,019	m 33,384	.125	4,173	m 7,674,556	.045	345,355	m 1,258,561	.057	71,738	889,481
1912	m 66,488	m 86,002	m 714,909	.615	439,669	m 23,885	.165	3,941	m 5,730,222	.045	257,860	m 308,681	.069	21,299	808,771
1913	m 56,763	m 50,282	m 805,343	.604	486,427	m 31,647	.155	4,905	m 3,398,364	.044	149,528	m 454,875	.056	25,473	716,615
1914	m 27,952	m 19,304	m 615,734	.588	340,501	m 32,586	.133	4,334	m 1,401,795	.039	54,670		.051		418,809
1915	28,071	m 33,039	m 291,807	.507	147,946	m 8,943	.175	1,565	m 2,382,128	.047	111,960	m 85,984	.124	10,662	305,172
1916	m 38,103	m 31,124	m 373,956	.658	246,063	m 13,138	.246	3,232	m 2,295,087	.069	158,361	m 240,575	.134	32,237	471,017
1917	m 32,755	m 10,101	m 361,517	.824	297,890	m 19,297	.273	5,268	m 1,305,744	.086	112,294	m 54,971	.102	5,607	431,160
1918	m 28,372	m 13,943	m 640,969	1.00	640,969	m 3,490	.247	862	m 989,620	.071	70,263		.091		726,027
1919	m 16,718	m 9,083	m 369,575	1.12	413,924	m 355	.186	66	m 934,113	.053	49,508	m 96,274	.073	7,028	479,609
1920	m 12,597	m 5,710	m 272,322	1.09	296,831	m 1,120	.184	206	m 531,537	.08	42,523		.081		345,270
		2,715,113	44,038,801		28,704,532	268,679		43,320	197,429,511		8,723,465	27,572,407		1,511,944	41,698,374

m U. S. Geol. Survey Mineral Resources.

BIBLIOGRAPHY.

' The literature on the geology and ore deposits of the San Juan region is extensive. Lists of the publications of the United States Geological Survey on mining districts in this region may be found in Bulletin 507, by J. M. Hill. Below are listed only those papers that treat the geology and ore deposits of the Creede district, which is separated from other areas of mineralization in the San Juan region by a fairly extensive area that is not known to contain mineral deposits. Besides the papers listed there are also numerous brief notes of timely interest in the technical and mining journals.

EMMONS, W. H., and LARSEN, E. S., A preliminary report on the geology and ore deposits of Creede, Colo.: U. S. Geol. Survey Bull. 530, pp. 42–65, 1913.

Discusses briefly the principal results of the United States Geological Survey investigations at Creede, of which the present paper is the final report.

EMMONS, W. H., and LARSEN, E. S., The hot springs and mineral deposits of Wagonwheel Gap: Econ. Geology, vol. 8, pp. 235–247, 1913.

Describes a fissure vein near which hot springs now issue.

FOSHAG, W. F., The crystallography and chemical composition of creedite: U. S. Nat. Mus. Proc., vol. 59, pp. 419–424, 1921.

HENDERSON, C. W., U. S. Geol. Survey Mineral Resources, 1908, pt. 1, p. 390, 1909.

Table of production and notes on development at Creede.

HENDERSON, C. W., idem, 1909, pt. 1, p. 320, 1910.

Table of production and notes on the mining conditions at Creede. Seven companies produced ore, the total tonnage of Mineral County being 64,941 tons. Over one-half of the product was milled, the remainder being shipped to smelters as crude ore.

HENDERSON, C. W., idem, 1910, pt. 1, p. 425, 1911.

The statistics of production in 1910 and mining conditions of the district are reviewed.

HENDERSON, C. W., idem, 1911, pt. 1, p. 551, 1912.

LARSEN, E. S. See Emmons and Larsen.

LARSEN, E. S., and WELLS, R. C., Some minerals from the fluorite-barite vein near Wagonwheel Gap, Colo.: Nat. Acad. Sci. Proc., vol. 2, p. 360, 1916.

LARSEN, E. S., and WHERRY, E. T., Halloysite from Colorado: Washington Acad. Sci. Jour., vol. 7, pp. 178–180, 1917.

LEE, H. K., Gases in metalliferous mines: Colorado Sci. Soc. Proc., vol. 7, pp. 163–188, 1903.

Discusses the geology and ore deposits of the Amethyst lode and describes the gas that issues from the walls into workings in the north end of the lode. An analysis of the gas is reported.

LINDGREN, WALDEMAR, and others, U. S. Geol. Survey Mineral Resources, 1905, p. 205, 1906.

Gives a table showing production of Mineral County and notes on mining conditions at Creede.

NARAMORE, CHESTER, U. S. Geol. Survey Mineral Resources, 1907, pt. 1, p. 262, 1908.

Notes on developments at Creede, with a table showing production.

PARMALEE, H. C., Zinc ore dressing in Colorado; the Creede district: Met. and Chem. Eng., Dec., 1910, pp. 677–678.

Describes the treatment of Creede ores at the concentrating plants of Creede. '

RICKARD, T. A., The development of Colorado's mining industry: Am. Inst. Min. Eng. Trans., vol. 16, pp. 834–848, 1896.

Gives an account of the early history of the Creede district and of its early production.

Chapter II.—OUTLINE OF THE GEOLOGY.

The bedrock exposed in the area shown on the map of Creede and vicinity and for a number of miles in all directions is made up entirely of Tertiary volcanic rocks and is a part of the great volcanic field of the San Juan Mountains. Lava flows form the greater part of the material, but there are a number of tuff and breccia deposits and a few small intrusive bodies. Although the rocks show little variety and with the exception of a single andesite formation are all classified as rhyolites and quartz latites, they belong to four distinct periods of eruption and are separated by very irregular surfaces of erosion. Furthermore, the rocks of each eruptive series consist of a number of subdivisions which are in turn commonly separated by irregular erosional surfaces.

In the course of the general survey of the San Juan Mountains, which has been in progress for some years, under the direction of Whitman Cross, the volcanic rocks have been studied and mapped over nearly the whole of the mountains, including the San Cristobal and Creede quadrangles. The lavas and associated lake beds and other clastic deposits of the Creede area, though in part local in extent and character, form a normal portion of the great San Juan sequence, and the two lower groups of eruptive rocks, here called Alboroto group and Piedra group, correspond to parts of the Potosi volcanic series. The lower part of the lower (Alboroto) group, including the Outlet Tunnel quartz latite, the Willow Creek rhyolite, and the Campbell Mountain rhyolite, differs somewhat from the normal rocks at this horizon, and as these rocks are not known at any great distance from Creede they probably represent a group of local flows.

The relations of the formations in the Creede district to those of neighboring areas on which reports have already been published are shown in the accompanying table.

AGE OF THE ROCKS.

The rocks are all believed to be of Miocene age, as plant remains collected under Bristol Head, to the west, from the Huerto formation, which lies between the Alboroto and Piedra groups, are closely related to those of the Florissant lake beds, which are Miocene and probably upper Miocene, and plant remains collected from the overlying Creede formation are also closely related to the Florissant flora.

POTOSI VOLCANIC SERIES.

ALBOROTO GROUP.

The oldest rocks exposed in the area comprise a succession of mapped formations consisting of rhyolites and quartz latites, chiefly in flows but containing some clastic material. These rocks collectively correspond to the Alboroto formation of the Potosi volcanic series as subdivided in Bulletin 13 of the Colorado Geological Survey, and are therefore here called Alboroto group. The lowest formation of this group, the Outlet Tunnel quartz latite, is exposed only in two small areas in the bed of East Willow Creek. (See Pl. II.) It is made up chiefly of biotite-hornblende-quartz latite but contains some pumiceous rhyolite; it comprises both flows and fragmental material. It is overlain irregularly by the Willow Creek rhyolite, which is made up of several flows of purple-drab* to gray fluidal banded felsitic rhyolite. Just north of Creede over 1,000 feet of this rhyolite is exposed without evidence of more than a single flow. The Campbell Mountain rhyolite overlies a somewhat irregular surface of the Willow Creek rhyolite and is made up of flows of a dull reddish-brown or drab mottled flow breccia. It attains a thickness of 1,000 feet near Creede but becomes much thinner to the northeast and east. The Phoenix Park quartz latite is chiefly above the Campbell Mountain rhyolite but is in part interbedded with that formation. It is commonly a light red-brown biotite-hornblende-quartz latite, chiefly in flows but containing some rather coarse breccia. It attains a great thickness at the head of East Willow Creek, but in the area shown on Plate II only a few hundred feet is exposed. The Equity quartz latite is closely related to the Phoenix Park quartz latite and probably represents a great flow in that series. It was recognized only in the northern part of the area in the drainage basin of West Willow Creek, where it overlies the Campbell Mountain rhyolite rather regularly and has a thickness of about 1,000 feet. It is in large part a single flow of a Quaker drab fluidal biotite-quartz latite.

PIEDRA GROUP.

The group of rocks composing the upper part of the Potosi volcanic series in this district corresponds to the Piedra formation of Bulletin 13 of the Colorado Geological Survey and is therefore here called Piedra group. These rocks have at their base a surface of great irregularity, and their extrusion was evidently preceded by a considerable period of erosion during which the streams cut canyons in the rocks of the Alboroto group comparable in depth and ruggedness to those of the present streams but in no wise related to the canyons of to-day.

The rocks of the Piedra group filled in these newly made canyons and probably covered nearly or quite all of the region. The lowest of the Piedra group of flows and clastic layers is a hornblende-quartz latite in thin flows and breccia beds, which is exposed only in the drainage basins of Rat and Miners creeks. In some places it is several hundred feet thick; elsewhere it is entirely absent. It is overlain by a few hundred feet of light red-brown rhyolite flow breccia and tuff (the Windy Gulch rhyolite breccia), which contains abundant included fragments of pumice. A tridymite latite, locally 400 feet thick, mostly in one great flow, overlies this rhyolite breccia rather irregularly. The rock is red brown and is characterized by prominent fluidal banding, by the presence of much tridymite in the more porous bands, and by abundant crystals of orthoclase, plagioclase, and biotite about 1 millimeter in diameter. The tridymite latite is overlain rather regularly by a series of andesites which varies greatly in thickness but nowhere exceeds 500 feet.

In the eastern part of the area the two lower formations of the Piedra group described in the preceding paragraph are lacking, and their place is in part taken by a great flow rhyolite, the Mammoth Mountain rhyolite, which is locally 1,000 feet thick. This rock is a red-brown flow breccia and is very similar to the Campbell Mountain rhyolite of the Alboroto group. It is overlain by several hundred feet of rhyolite tuff with associated thin flows, and those beds in turn are overlain by the tridymite latite to the east of the area mapped.

A succession of quartz latites overlies rather irregularly the andesite or tridymite latite and locally rests on other rocks. These quartz latites carry rather abundant crystals of plagioclase, quartz, orthoclase, biotite, augite, hornblende, and titanite about 1 millimeter in diameter. The lowest formation of this succession consists of several hundred feet of nearly white quartz latite tuff with some interbedded flows. The intermediate formation, the Rat Creek quartz latite, is made up in large part of flows but contains some tuff. The upper formation, the Nelson Mountain quartz latite, is a persistent mesa-forming flow, about 200 feet thick.

CREEDE FORMATION.

The Creede formation was deposited in a lake that occupied a valley carved out of the rocks of the Potosi volcanic series. This valley was deeper than that of the present valley of the Rio Grande and occupied about the same position from Antelope Park to Wagonwheel Gap.

The lower part of the Creede is made up chiefly of fine-textured thin-bedded rhyolite tuffs, with some coarser material, especially near the borders of the old lake. It contains numerous bodies of travertine, which indicate the presence of abundant hot springs during this period. The upper part of the Creede formation is of somewhat coarser texture and consists chiefly of fairly well bedded breccia and conglomerate with some fine tuff and intercalated thin flows of soda rhyolite.

FISHER QUARTZ LATITE.[3]

A later series of lava flows overlies the older rocks very irregularly and in the area described in this report is made up of a single great flow, the Fisher quartz latite, which is characterized by abundant large crystals of plagioclase, biotite, and augite. To the north and east this formation embraces a great thickness of flows and tuff breccias made up of quartz latites and related rocks which show numerous large phenocrysts.

INTRUSIVE ROCKS.

Intrusive rocks are present in this area only in a few comparatively small bodies. Four types have been recognized. The oldest is a rhyolite porphyry, which was intruded as irregular or sill-like bodies into the rocks of the lower or Alboroto group of the Potosi volcanic series. It is a nearly white rock containing large crystals of glassy orthoclase. A single small dike of andesite cuts the rocks of the upper or Piedra group of the Potosi rocks west of Rat Creek. In Miners Creek a number of dikes of quartz latite porphyry cut the Piedra rocks and are probably related to the flows of the Fisher quartz latite. The rock of these dikes is dense and carries rather abundant large crystals of plagioclase and some of biotite, augite, and hornblende. A hornblende-quartz latite porphyry is present in some of the mine workings.

QUATERNARY DEPOSITS.

After the volcanic activity had ceased erosion again became the dominant geologic agent, and the present mountains and canyons were carved from the great volcanic pile. There is reason to believe that the erosion took place in two stages. During the first stage erosion proceeded to moderate maturity with the main stream channel about 1,000 feet higher than that of the present Rio Grande and developed the broad valleys and comparatively gentle rolling hills of the basin of upper Windy Creek and of the other streams of the area. The second stage opened with a considerable increase in the gradient of the streams. This began in the larger streams and worked upstream, increasing the gradient and developing the deep, rugged canyon above Creede and Sunnyside. After the streams had cut down their beds nearly to their present level, the upper parts of the main streams were occupied by glaciers that widened the

---

[3] Used by H. B. Patton, quoting Whitman Cross and E. S. Larsen, in Colorado Geol. Survey Bull. 13, p. 20, 1917, and defined by Patton on pp. 23–33.

valleys but did not cut them much deeper.   Since the disappearance of the glacial ice the streams have deepened their channels but little.

The terminal moraines of the glaciers of Rat Creek and both forks of Willow Creek are within the area included in this report.   Large landslides are conspicuous geologic features, and talus, terrace gravels, and alluvium cover small areas.

GEOLOGIC STRUCTURE.

The chief deformation of the region has been rather complex block faulting, with some tilting near the faults.   Nearly all the ore deposits lie along these faults.   The faults are believed to be later than any of the volcanic rocks of the area, and they are earlier than the development of the present topography.   The area has also probably been gently tilted toward the south.

Chapter III.—ROCKS OF THE ALBOROTO GROUP.

GENERAL FEATURES.

The oldest rocks exposed within the area covered by the map of Creede and vicinity (Pl. II) are a thick series of rhyolites and quartz latites belonging to the Alboroto group of the Potosi volcanic series. Only a few miles to the west, under Bristol Head, they are underlain irregularly by a series of andesitic rocks, and still farther up the Rio Grande, near the mouth of Lost Trail Creek, they overlie the San Juan tuff.[4] In other places, as in upper Ute Creek, west of Creede, and in Cebolla Creek, north of Creede, they or younger volcanic rocks directly overlie pre-Cambrian granites, schists, gneisses, quartzites, and other rocks; in still other places, as near Pagosa Springs and south of Gunnison River, they overlie Paleozoic and Mesozoic sediments.

The top of this group is likewise a surface of marked irregularity, owing to the considerable period of erosion that immediately preceded the extrusion of the rocks of the overlying Piedra group. A few miles to the west a great thickness of andesitic rocks lies between these two groups and beneath the erosional surface. In the Creede area, however, the rocks of the Piedra group directly overlie those of the Alboroto group, the intervening Huerto formation being absent.

The rocks of the Alboroto group as exposed near Creede are all highly siliceous; the greater part are rhyolites, although quartz latites are present in the northern part of the area and attain a great development farther north, on both forks of Willow Creek. The rhyolites and quartz latites alternate, and it is probable that they originated from different vents. The group has been subdivided into five formations—two rhyolites and three quartz latites—and most of these consist of several flows or of flows and associated tuffs. The lowest quartz latite, the Outlet Tunnel quartz latite, underlies at least a part of the Willow Creek rhyolite. Its upper surface is very irregular, and it may represent a much older period of eruption than the overlying flows. The Campbell Mountain rhyolite, wherever exposed, immediately overlies a rather irregular surface of the Willow Creek rhyolite. The second quartz latite, the Phoenix Park quartz latite, in general overlies the Campbell Mountain rhyolite but in places is interbedded with it. The Equity quartz latite, where recognized, immediately overlies the Campbell Mountain rhyolite, and it is believed to be more closely related to the Phoenix Park quartz latite than to the rhyolites.

---

[4] Cross, Whitman, U. S. Geol. Survey Geol. Atlas, Silverton folio (No. 120), p. 7, 1905.

OUTLET TUNNEL QUARTZ LATITE.

GENERAL CHARACTER AND OCCURRENCE.

The Outlet Tunnel quartz latite, which is the oldest rock exposed in the area mapped, is a chaotic aggregate of lava flows and breccia beds. It was found in only two small bodies in the canyon of East Willow Creek a short distance north of the Ridge mine and has not been recognized in the reconnaissance of the adjoining region. In the lower body it shows fair outcrops, although the contacts are everywhere covered so that its extent and form can be only roughly determined. The upper body is exposed only in the outlet tunnel, from which this rock receives its name. Talus from the overlying rhyolite covers the lower slopes of the hills, and the two bodies may be connected on the west side of the creek beneath the talus, but it appears more probable that the Willow Creek rhyolite separates the two bodies, and this interpretation is shown on the map. Only about 250 to 300 feet of this quartz latite is exposed, but as the base has not been reached it is not improbable that the part exposed represents only the top of a body of considerable thickness. It is overlain irregularly by the Willow Creek rhyolite, although the contact has nowhere been seen.

PETROGRAPHY.

Megascopic features.—The rocks of the flows and breccia fragments are chiefly biotite-hornblende-quartz latites of purple-drab to gray color.[5] They carry phenocrysts from 1 to 2 millimeters across of white plagioclase, commonly altered, glassy orthoclase, quartz, biotite, and altered hornblende, in an aphanitic groundmass; a few show in addition an occasional crystal of orthoclase or microperthite as much as 3 centimeters across. The groundmass equals or exceeds the phenocrysts in amount. The rocks are mostly rather dense, but some carry visible pores and gas cavities. Fluidal texture is present in much of the rock but is rarely conspicuous. Small inclusions are abundant in some of the rocks and consist chiefly of quartz latite of darker color than the host and carrying fewer phenocrysts; a few consist of rhyolite or andesite.

---

[5] As the color is rather characteristic of many of the rocks of the Creede district and as slight differences in color are among the most easily recognized differences between some of the rocks, accurate color descriptions of the rocks are highly desirable. Accurate color names require a standard of color nomenclature, and no such standard that is both sufficiently comprehensive and generally accepted exists. Robert Ridgway's "Color standards and nomenclature," published in 1912, is the best standard that we have, but to use it properly requires an actual comparison with the plates in Ridgway's book. It therefore seemed best to the writers to use in the text color names that could be properly understood by reference to a good dictionary and to include the more precise name according to Ridgway's color standards in footnotes. According to Ridgway's color standards the rocks of the Outlet Tunnel quartz latite are commonly light purple-drab (1''''b) to purple-drab (1''''), less commonly purplish vinaceous (1'''b), pallid Quaker drab (1''''''f), or pale neutral gray (d).

Microscopic features.—The microscopic study of the thin sections showed that about a third of most of these rocks is made up of phenocrysts of plagioclase, orthoclase, quartz, biotite, green hornblende, magnetite, apatite, titanite, and zircon, stated in the order of their abundance as roughly estimated. The plagioclases are much altered to calcite and sericite, in some specimens to kaolinite; the orthoclase is fresh; the quartz is greatly embayed from magmatic resorption; the apatite crystals have colorless ends and a faintly pleochroic, smoky yellow-brown core. The phenocrysts are embedded in a matrix which is indistinctly polarizing in specks and shreds, less commonly in minute rounded areas. It is clouded and carries numerous minute reddish shreds which are probably hematite.

Less common rock types.—Rare fragments in the breccia show more conspicuous and abundant crystals from 3 to 5 millimeters across of biotite and of plagioclase having the composition of andesine. The groundmass is uneven in size of grain and much coarser than that in the other rocks; it is granophyric or micrographic in texture.

Within the quartz latite breccia are thin, irregular bodies of rhyolite flow breccia. This rock is pale purplish gray [6] and shows a few crystals of quartz, biotite, and glassy orthoclase about a millimeter long in a glassy to aphanitic groundmass. It is characteristically porous and carries very abundant and prominent irregular-shaped inclusions of fibrous pumice as much as several centimeters across. Weathering removes the pumice fragments, leaving large ragged cavities. This rock closely resembles the Windy Gulch rhyolite breccia.

WEATHERING AND OUTCROPS.

The rocks are considerably altered through hydrothermal action. The plagioclase is sericitized, and the hornblende is altered to a crumbly dark-red material; secondary calcite and chlorite are abundant, and some sulphides are present. This rock is less resistant to erosion than the overlying rhyolite and gives inconspicuous outcrops that are largely masked by talus from the cliffs of the overlying rhyolite.

WILLOW CREEK RHYOLITE.

GENERAL CHARACTER AND DISTRIBUTION.

The thick series of flows of fluidal felsitic rhyolites characteristically exposed above Creede in the canyons of both forks of Willow Creek is here called the Willow Creek rhyolite. It is also prominent in Dry Gulch, in Miners Creek, and above the Equity mine in upper West Willow Creek. Its distribution is shown on Plate II. From reconnaissance work about Creede it is believed to come out from under the overlying rocks to the north at Bondholder; to the east it

---

[6] Ridgway's pallid purplish gray (67′′′′′*f*).

crops out more or less continuously for some miles, but has not been recognized east of Wagonwheel Gap; to the west it wedges out rapidly and is not present south of Bristol Head.

On the west slopes of Mammoth Mountain nearly 2,000 feet of this formation is exposed and the base is not seen; in other areas along both forks of Willow Creek there is nearly as great a thickness. Where its base is exposed on East Willow Creek above the Ridge mine it overlies an irregular surface of the Outlet Tunnel quartz latite and is not more than 200 feet thick. It therefore decreases in thickness from nearly 2,000 feet to about 200 feet in a distance of 1½ miles. It is possible that the Outlet Tunnel quartz latite represents a lens between flows of the Willow Creek rhyolite, but this is not believed probable.

PETROGRAPHY.

Megascopic features.—The rocks of these flows are commonly light to dark purple-drab, less commonly light drab, buff, or gray.[7] Fluidal structure is always present and is commonly prominent and characteristic. The main part of the rock is dense and has about the luster of freshly broken porcelain, but streaks or lenses are decidedly porous and have a paler color. These streaks as seen on plates broken along the banding are a few centimeters wide and several decimeters long; they are generally not over a few millimeters thick. They may form as much as 10 per cent of the rock. In addition to these larger streaks and grading into them are closely spaced narrow bands, giving the rock a beautiful fluidal structure. In places, notably on Rat and Miners creeks, the rock has a more delicate fluidal structure and a somewhat paler color. Near Weaver the rock breaks into thin plates along its well-developed and closely spaced fluidal banding; at other places it has an inconspicuous fluidal structure. In a few places it carries large cavities lined with drusy quartz crystals and partly filled with chlorite; rarely it carries inclusions of foreign rock.

The rock contains a very few visible crystals, mostly of orthoclase but in part of white plagioclase, commonly kaolinized, and a very little biotite. Otherwise it is felsitic or aphanitic—that is, its composition can not be determined in the hand specimen. A number of flows are probably present, but their close similarity and the lack of glassy layers or other recognizable distinctions at the tops or bottoms of flows makes it difficult to distinguish between them. Where the greatest thickness is exposed, on Willow Creek, if more than one flow is present they are almost identical in character and could not be distinguished. Near Sunnyside several flows, some of

---

[7] Commonly Ridgway's purple-drab, ranging from pallid to dark (1‴f to 1⁗i); less commonly purplish vinaceous (1‴b), light cinnamon-drab (13⁗b), a number of pallid tints of buff, lilac, drab, and gray, and the lighter tints of gull-gray.

them different from the rock of Willow Creek, have been included in this formation.

Microscopic features.—The rock is uniformly holocrystalline and contains few phenocrysts, chiefly of orthoclase, with some plagioclase, a very little biotite, and accessory apatite, iron ore, and zircon. The plagioclases are albite and albite-oligoclase and are commonly kaolinized. The orthoclase crystals are commonly broken and in sóme specimens show on their borders between crossed nicols a lacework that grades into the groundmass and is due to a growth of material from the groundmass on the crystals. The groundmass is beautifully banded. The main part is indistinctly polarizing, even with the highest magnification, and is clouded from minute bodies of ferritic material; numerous lenses or streaks are much coarser and are clear. These coarser streaks correspond to the lighter-colored bands and lenses seen in the hand specimens and make up a small percentage of the rock; they range in width from a few millimeters to a small fraction of a millimeter.

The larger lenses show a marked banding and concentration of quartz in their interior. These bands, of which there are commonly three or four in a lens, are not sharply bounded but grade into one another. The outer band, which is not always present, is a few tenths of a millimeter across and is made up of parallel fibers which project from the walls and extend across the band, resembling delicate spherulites. Sharply bounded from or grading into this fibrous band or forming the outer band of some of the lenses is a band made up of comparatively coarsely crystallized quartz and orthoclase. The orthoclase is chiefly in fairly well-formed elongated crystals, a few of which are greatly elongated or acicular. In part they appear to have grown from the walls and commonly project from the walls; in part they are embedded in the quartz crystals poikilitically. They vary greatly in size, and the largest are 0.3 millimeter long. The quartz is interstitial and in a few places is crystallographically continuous with the quartz grains of the interior. As quartz becomes more abundant toward the interior and both quartz and orthoclase become more coarsely crystalline, this band grades into the next, which is made up of an aggregate of interlocking quartz grains, commonly 1 millimeter across, with some gas cavities. Some bands have a core which is largely pore space and into which project drusy crystals of quartz and orthoclase that can be easily seen with a pocket lens. Tridymite was not observed.

Plates V–VII are photomicrographs of thin sections of the Willow Creek rhyolite showing these coarsely crystalline lenses. Plate V shows two lenses with the normal groundmass between; Plate VI, *A*, shows in detail one of the lenses and the normal groundmass on both sides; Plate VI, *B*, shows between crossed nicols the same area as Plate VI, *A;* Plate VII shows a part of one of the broader lenses. Some of the very broad lenses do not show an interior quartz band, but its place is taken by irregular areas of quartz grains which are scattered through these lenses, as shown on Plate VII. The orthoclase is in general irregularly distributed. The smaller lenses do not show so distinct a banding, and some of the narrower streaks are mere strings of crystals. In some specimens these coarsely crystalline lenses are spherulitic in crystallization or micrographic, with little if any concentration of quartz in the central part. These lenses are less clouded than the main groundmass, as the ferritic material is collected in small grains and shreds which are in part black and opaque, in part reddish brown and translucent. In addition there are a few minute colorless grains with a rather high index of refraction and without perceptible birefringence. The orthoclase is somewhat clouded, but the quartz is clear except for scattered streaks and irregular areas that carry abundant minute gas or liquid inclusions. No liquid inclusions with gas bubbles were found. Phenocrysts of orthoclase and plagioclase are present in some of the larger lenses.

These lenses, which differ from the main body of the flow in their more porous character, coarser crystallization, euhedral form of the orthoclase, and banded structure, evidently represent the result of magmatic segregation before the magma came to rest. Relief of pressure due to the extrusion of the lava might have caused the mineralizers to concentrate into rounded bodies, perhaps in part as bubbles or aggregates of bubbles sealed in the lava, in part in solution in the magma; as the viscous lava flowed these bodies were drawn out into their present forms. The presence of the mineralizers held back the crystallization of these streaks until after the main body had solidified; it also caused the coarser crystallization, the concentration of the quartz in the center, and the presence of the gas cavities. On account of the coarseness of the crystallization, the euhedral development of the orthoclase, and the concentration of the orthoclase on the walls, the writers believe that these streaks were not highly viscous at the time of their crystallization. The crystallization of the lava and especially of the coarser lenses is believed to have depended more on the loss of the mineralizers than on the cooling of the magma. These mineralizers, released during the crystallization of the magma, may have caused the kaolinization of the plagioclase, which is almost universal and does not appear to be due to weathering.

These streaks show considerable resemblance to some veins of adularia. The concentration of orthoclase on the borders and the interlocking of the quartz grains are characteristic of both. However, the form of the orthoclase in the veins is rhomboidal, whereas in the streaks of the rhyolite it is characteristically prismatic. The streaks also show a resemblance to the small, irregular pegmatitic bodies which are common in some granites.

CHEMICAL COMPOSITION.

Three chemical analyses of this rock, made by W. C. Wheeler in the laboratory of the United States Geological Survey, are given below.

Chemical analyses of the Willow Creek rhyolite.

	1	2	3		1	2	3
SiO_2	73.53	76.26	77.36	TiO_2	0.19	0.15	0.16
Al_2O_3	12.87	11.30	11.37	CO_2	.23	.19	.06
Fe_2O_3	.88	.52	.31	P_2O_5	Trace.	.01	.03
FeO	.64	.34	.36	S	.02	.26	.33
MgO	.56	.02	.14	Cr_2O_3	None.		
CaO	.07	.23	.30	MnO	.09	.05	.03
Na_2O	.63	2.81	1.38	BaO	.05	.49	.05
K_2O	8.92	6.77	7.28				
H_2O-	.40	.39	.55		99.87	99.93	99.97
H_2O+	.70	.14	.26				

1. Typical fluidal rhyolite from Solomon adit.
2. Typical fluidal rhyolite.
3. Typical fluidal rhyolite from Bachelor shaft, Nelson adit, about 30 feet from vein.

The norms of these analyses, calculated in accordance with the quantitative system of Cross, Iddings, Pirsson, and Washington,[a] are as follows:

Norms of Willow Creek rhyolite.

	1	2	3		1	2	3
Quartz	34.7	34.8	40.1	Wollastonite		0.5	
Corundum	2.0	....	1.5	Diopside		.7	
Orthoclase	52.8	40.0	43.4	Pyrite		.5	0.6
Albite	5.2	20.4	12.0	Magnetite	1.4		.5
Anorthite	.3	....	1.4	Ilmenite	.3	.3	.3
Hypersthene	1.7	....	.5	H_2O, CO_2, etc	1.4	.7	1.2
Acmite		1.4					
Na_2SiO_3		.4			99.8	99.7	100.5

The rock represented by analysis 1 falls in Class I, order 3 near order 4, rang 1, and subrang 1—I.''4.1.1(2)—and is letachose. Chemically this rock is remarkable for its high silica and potash, low soda, and very low lime. The rock of analysis 2 falls in Class I, order 4 near order 3, rang 1, and subrang 2—I.(3)4.1.2(3)—and is an omeose. It differs from No. 1 chiefly in the greater amount of soda; it is remarkable that the BaO in this rock exceeds the CaO; this may be due to the presence of a small amount of secondary barite. The rock of analysis 3 is a magdeburgose (I.3''.1.2) and is higher in silica than the other two, but is otherwise intermediate between them. In all three rocks the norm differs from the mode

---

[a] Quantitative classification of igneous rocks, Chicago and London, 1903.

chiefly in the absence of corundum in the mode and the presence of biotite instead of pyroxene.

The analyses differ somewhat more than would be expected from specimens collected from the same flow. Analysis 1 shows lower silica and especially lower soda and higher potash. This may not represent an original difference in the rocks but may indicate that the rock represented by analysis 1 has had its plagioclase altered to kaolinite. The considerable amount of water shown in the analyses and the presence of corundum in the norms bear out this suggestion. It is possible also that adularia has been added.

STRUCTURE.

The strike and dip of the fluidal banding of the rock varies from place to place in an irregular manner. The dip is commonly steep, and so far as could be observed this is not due to tilting subsequent to the extravasation of the lava but to the irregular flow of a viscous magma probably over an uneven surface. A nearly vertical sheeting due to crushing or perhaps in part to shrinkage on cooling is commonly present, and locally it is close spaced and prominent. Where two or more such systems of sheeting are developed the rock breaks into rude columns or pencils. Plate X, B, shows the details of a typical outcrop; it pictures the cliffs about the Monte Carlo mine, on the southeast slope of Campbell Mountain. The broken cliffs above the talus are about 1,000 feet high.

WEATHERING AND ALTERATION.

The weathering of this rock has been largely mechanical and due to frost action and differential expansion caused by changes in temperature, which are rather extreme in this high altitude. Gravity has also been an important factor, and rock slides of various sizes are very common. The prominent fluidal structure and the sheeting have greatly influenced the breaking down of the rock. Soil is scant, and great piles of small rock fragments, made up largely of small plates of nearly fresh rock, are characteristic at the base of the cliffs. The gentler slopes of the highlands show a scant soil with abundant fragments of nearly fresh rock scattered through it.

Near the mineral veins or other sulphide bodies the rock has been more or less altered by the mineral solutions. In places near the surface along the Amethyst vein there are considerable bodies of white, kaolinized rock, probably due to leaching by acid solutions. In addition, much of the rock, though showing no other signs of alteration, has its phenocrysts of plagioclase altered to kaolin. The kaolinization evidently had no relation to the zone of weathering or to zones of mineralization and is believed to have taken place immediately after the solidification of the lavas and before complete cooling.

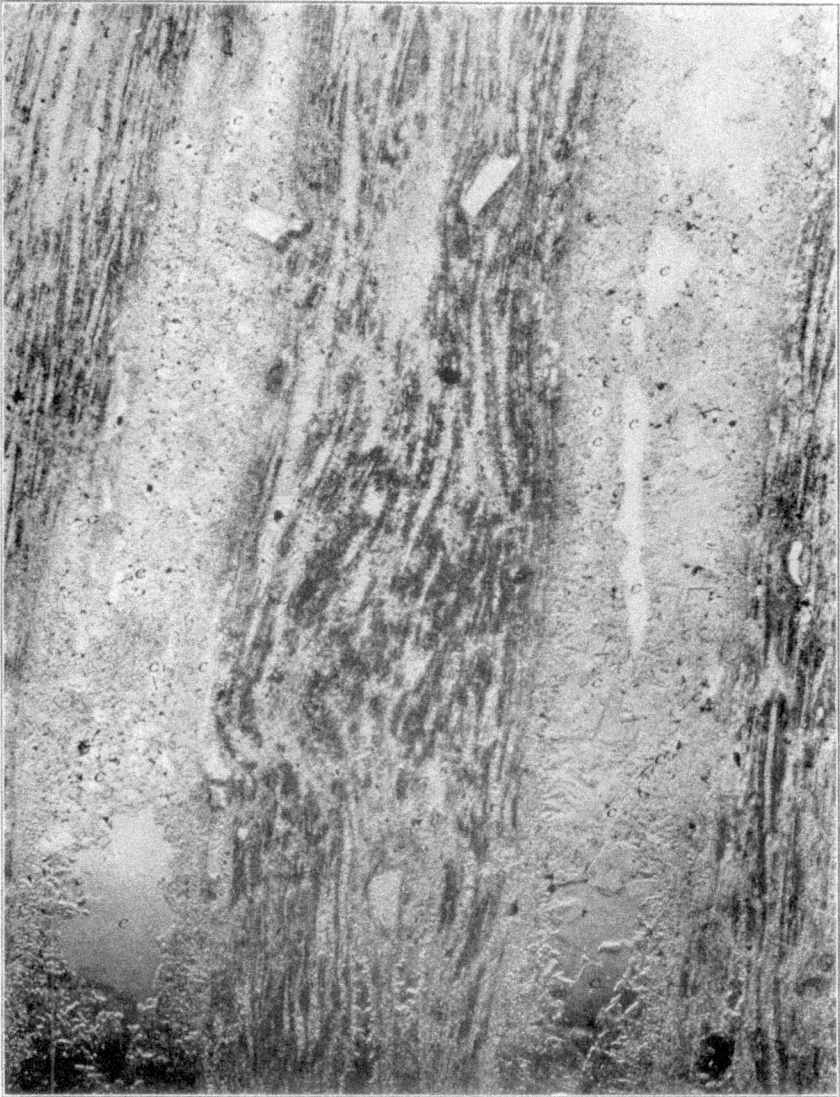

PHOTOMICROGRAPH OF WILLOW CREEK RHYOLITE.

Cross section of two broad layers of coarsely crystalline quartz and orthoclase. The orthoclase stands out in relief. c, Gas cavity. (See text.) Inclined illumination to bring out relief. Magnified 50 diameters. Details of one of the bands are shown in Plates VI and VII.

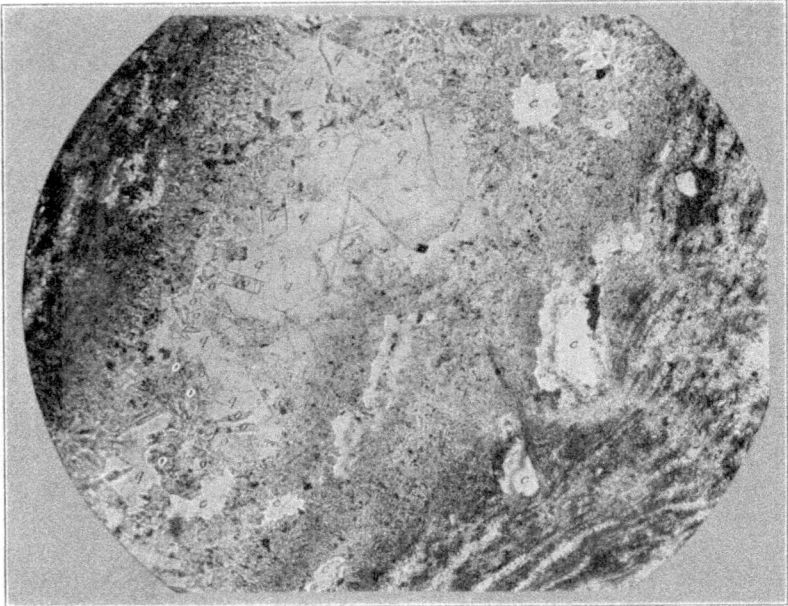

A. COARSELY CRYSTALLINE LAYER WITH VERY FINE TEXTURED GROUND-
MASS ON BOTH SIDES.

The orthoclase crystals (*o*) stand out in relief against the quartz (*q*). *c*, Hole in section.
Shows change in crystallization from border to center. (See text.) Magnified 100
diameters.   Inclined illumination to bring out relief.

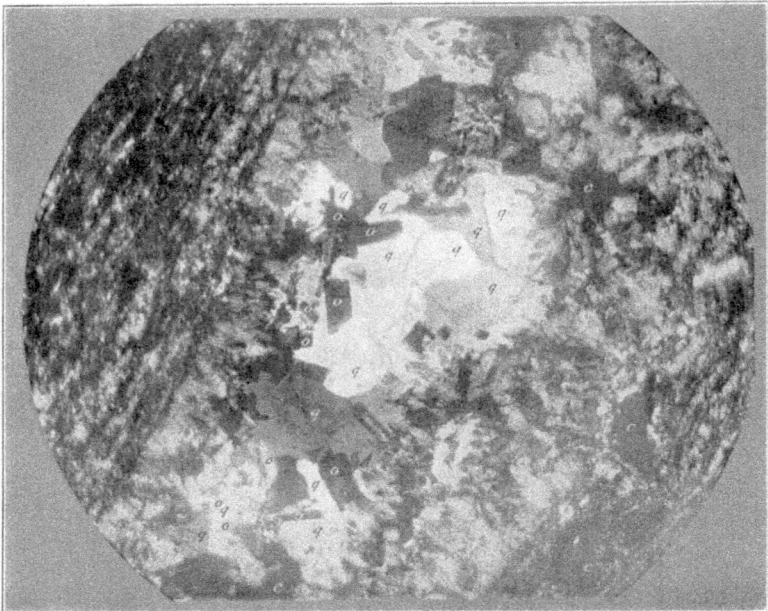

B. AREA SHOWN IN *A*, WITH CROSSED NICOLS.

q, Quartz; *o*, orthoclase; *c*, cavity. (See text.)   Magnified 100 diameters.

PHOTOMICROGRAPHS OF WILLOW CREEK RHYOLITE.

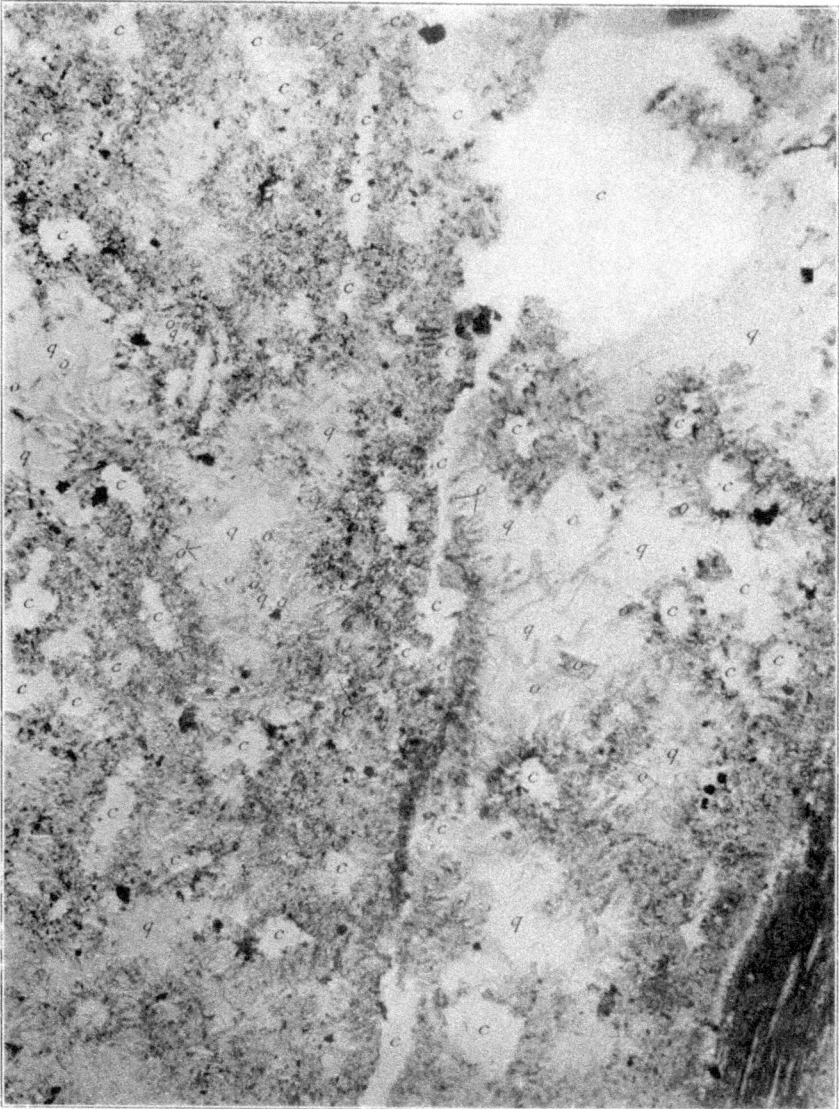

PHOTOMICROGRAPH OF WILLOW CREEK RHYOLITE.

Section across part of one of the broad layers of coarse crystallization. The orthoclase crystals (o) stand out in relief against the quartz (q). c, Cavity. The contact with the very finely crystalline groundmass is shown in the lower right-hand corner. (See text.) Magnified 100 diameters.   Inclined illumination to bring out relief.

A. EQUITY FAULT.

B. WILLOW CREEK CANYON ABOVE CREEDE.

CLIFFS OF WILLOW CREEK RHYOLITE NEAR FORKS OF WILLOW CREEK.

A. CLIFFS OF WILLOW CREEK RHYOLITE NEAR FORKS OF WILLOW CREEK.

B. SHEETING OF WILLOW CREEK RHYOLITE.

TOPOGRAPHY AND SCENERY.

The rock commonly crops out in jagged and more or less broken cliffs with great talus heaps at their bases. Of all the rocks in the area it is exceeded in hardness and resistance to weathering only by the latite of MacKenzie Mountain, and most of the rugged topography and deep canyons about Creede are in this rock. Wherever the streams have cut deeply into it they have sharp canyons or gorges with steep, rugged walls and broken cliffs for 1,000 feet or more. The uniformity of the material does not lead to the development of benches. Plates VIII, B, IX, and X show characteristic outcrops of this rock. Plates IX and X, A, are reproductions of photographs looking up the creek from a point a few hundred yards below the forks of Willow Creek. The cliffs are all of the Willow Creek rhyolite; that in the center between the forks of Willow Creek is about 1,500 feet high. The talus slopes at the bases of the cliffs are at about the top of the Willow Creek rhyolite, and the timbered slopes on the right, above the shoulder, show the characteristic topography of the Campbell Mountain rhyolite. Plate X, B, shows a more detailed view of these cliffs. It was taken from a point about 500 feet below the Mollie S. mine and shows the cliffs on the west side of East Willow Creek. These cliffs are about 1,500 feet high. The Monte Carlo mine building is indistinctly shown near the top of the cliffs to the right of the center. Plate VIII, B, shows the canyon of Willow Creek above Creede, which is cut in this rock. These rugged outcrops are especially well developed on both forks of Willow Creek and in Dry Gulch. On Miners Creek they are nearly as characteristic.

CAMPBELL MOUNTAIN RHYOLITE.

GENERAL CHARACTER AND DISTRIBUTION.

Overlying the Willow Creek rhyolite rather irregularly is a rhyolite low breccia which is here called the Campbell Mountain rhyolite. In most places no evidence was seen of more than one flow, but on East Willow Creek two flows of this type are separated by a few hundred feet of the Phoenix Park quartz latite. The Campbell Mountain rhyolite is present on both sides of Miners Creek, in the lower part of the Rat Creek drainage basin and northeast of Monon Hill. It was also found in Windy Gulch just south of Bachelor. Several isolated outcrops, partly bounded by faults, are present on both sides of Willow Creek, a short distance above Creede. It caps parts of Mammoth Mountain. A narrow band extends from the ridge north of Mammoth Mountain in a northwesterly direction, crosses East Willow Creek at Phoenix Park, and continues southwestward to Campbell Mountain. Just northwest of this locality the Campbell Mountain rhyolite is cut out for a short distance by the Mammoth Mountain rhyolite, but it reappears as a thicker member on Nelson Creek and to the northwest. It has been recognized on both sides of the Equity fault. It is not exposed to the west of West Willow Creek except near its mouth, but in the underground workings along the Amethyst vein this rhyolite forms the hanging wall. The details of its distribution in the Creede area are shown on Plate II. To the west of this area it is well developed in the drainage basin of Shallow Creek but is not present south of Bristol Head. It has been recognized as far east as Wagonwheel Gap and as far north as Bondholder, on Spring Creek. It has not been recognized south of the Creede area.

About a quarter of a mile a few degrees south of east of the Captive Inca shaft, in a great landslide area at the face of a short tunnel, is an exposure, believed to be in place, of a somewhat altered rhyolite flow breccia or tuff which probably belongs to this rhyolite, although it has more of the appearance of some of the rocks of the Windy Gulch rhyolite breccia. The Windy Gulch rhyolite is not known to occur in this part of the area and if the exposure does not belong to the Campbell Mountain rhyolite it is probably a part of the overlying tuff and if so may be a large block of slide rock.

The Campbell Mountain rhyolite varies greatly in thickness, for three reasons—it flowed over a rather uneven surface of the underlying rhyolite; in most places erosion had cut deeply into it before the overlying rocks of the Piedra group were extruded; it thins out and the flows wedge out between flows of the Phoenix Park latite in the northeast corner of the area. The maximum thickness is shown south of Nelson Mountain, where nearly 1,000 feet is present and originally it was probably thicker as the top is a surface of erosion. Less than a mile to the south it is entirely eroded, and the rocks of the Piedra group rest directly on the Willow Creek rhyolite. To the west of the Creede area, on Shallow Creek, the thickness is probably even greater; although a few miles farther west, south of Bristol Head, this flow wedges out and the Alboroto group is made up entirely of quartz latite.

CONTACTS.

The basal contact of the Campbell Mountain rhyolite with the Willow Creek rhyolite, where observed, is commonly sharp and shows evidence of some erosion preceding the extrusion. West of Willow Creek, above Creede, at the Exchequer mine, the contact is well exposed, especially in the mine workings. It dips steeply to the west. The two rhyolites, each in typical development, show a sharp contact and are closely adherent. The Campbell Mountain rhyolite has plucked off numerous blocks from the underlying Willow Creek rhyolite and carries them as angular inclusions as much as 1 foot across. The Campbell Mountain rhyolite at the contact is not noticeably different from the normal rock except for the presence of the numerous rather large included blocks of the underlying Willow Creek rhyolite, and these extend for only a few feet from the contact. At other places the two rocks appear to grade into each other, and even with continuous exposures there is as much as 100 feet of rock of intermediate character between the two typical rocks, as is well shown on the east side of East Willow Creek about half a mile above the Ridge mine. Along Miners and Rat creeks also some difficulty was experienced in locating the contact, partly on account of the lack of a sharp contact and distinctive character in the two rhyolites and partly on account of poor exposures.

The upper contact of the Campbell Mountain rhyolite is everywhere sharp, but some of the overlying rocks so closely resemble it that the separation was somewhat difficult.

PETROGRAPHY.

Megascopic features.—The Campbell Mountain rhyolite is commonly reddish brown or drab [9] and has a rather dull luster, a delicate, indistinct fluidal texture, and a characteristic mottled or spotted appearance due to inclusions of lighter and darker shade. It is generally porous, with pores less than 1 millimeter across; the larger ones are lined with drusy crystals of quartz and feldspar or are partly filled with a network of minute crystals arranged in strings like rock candy. The inclusions are in part angular fragments of a somewhat darker quartz latite similar to the Outlet Tunnel quartz latite; few of them are 1 centimeter across, and nearly all are much altered. Fragments of the underlying Willow Creek rhyolite are abundant only near the contact with that formation; fragments of andesite and other rocks are present here and there. Even more abundant and characteristic are the rounded to angular fragments of rhyolite, which differ from the host chiefly in their more porous character, slightly paler color, and coarser crystallization. Some of them have irregular serrated outlines; some are flattened parallel to the flow structure or rudely lenticular. They are commonly very ragged and as seen in the specimens fray out at their ends. They are believed to represent, in part at least, products of magmatic segregation not very different from the lighter lenses in the Willow Creek rhyolite, but they may represent fragments of the magma which had already crystallized and were torn from the sides of the vent by the viscous lava. The inclusions of both kinds are commonly bordered by narrow bands of somewhat paler tint than the main rhyolite mass.

---

[9] Ridgway's livid brown (1'''), purple-drab (1''''), light purple-drab (1''''b), cinnamon-drab (13''''), or light Quaker drab (1'''''b).

A rock of somewhat different appearance is prominent in the body south of Nelson Mountain and is also present near Sunnyside and in other places. This rock is light gray to white [10] and has a delicate flow structure. It carries numerous included flat, splinter-like fragments of a white, rather porous rhyolite, in addition to some of quartz latite and the Willow Creek rhyolite; in most of these fragments the maximum diameter is parallel to the flow structure. It differs from the drab type chiefly in its color, but it also has a more prominent fluidal structure and somewhat more conspicuous and abundant included fragments, which are commonly porous. It is not believed to be a distinct flow, but rather the top or other part of one or more flows of the normal rock. It is everywhere closely associated with the normal rock, and south of Nelson Mountain it makes up most of the upper part of the body, although it is not confined to any particular horizon. In places rocks transitional between this and the normal rock were found. Locally leaching of the typical rock has given rise to a rock similar to the gray type, and much of this type may have been thus developed.

Both types show a few crystals 1 millimeter across of glassy orthoclase, white plagioclase, biotite, and quartz in an aphanitic groundmass. Glassy or highly vesicular layers have not been recognized.

Microscopic features.—The thin sections show that the rocks differ but little from those of the underlying Willow Creek rhyolite; on the whole they have somewhat more abundant and more calcic plagioclase, some quartz phenocrysts, and a slightly different groundmass, but individual sections can not be easily distinguished. They carry scattered phenocrysts, chiefly of orthoclase, with some calcic andesine, embayed quartz, and biotite, and accessory apatite, iron oxide, and zircon. The groundmass is made up of quartz and orthoclase and is delicately fluidal, most of it very finely crystalline to submicroscopic, with streaks and irregular patches of coarser spherulitic or granophyric crystallization. The blotches so prominent in the hand specimens are sharply bounded from the host and are somewhat coarser in crystallization but otherwise similar.

WEATHERING AND OUTCROPS.

This rock breaks up readily under the influence of weathering into fragments that are flattened parallel to the flow banding and commonly several inches across; it gives poor outcrops and commonly forms gentle slopes strewn with fragments of the underlying rock. The rock is much less resistant than the underlying Willow Creek rhyolite, and to this difference is due much of the variety in topography and scenery on both sides of Willow Creek above Creede.

---

[10] Ridgway's gull-gray (*d* [8]), light gull-gray (*f* [9]), pale mouse-gray (15′′′′′*d*), or white.

The cliffs consist almost entirely of the Willow Creek rhyolite, and the soft, gentle slopes above most of the benches and the flat tops of the ridges are commonly formed from the Campbell Mountain rhyolite. The marked difference between the outcrops of the two formations aids greatly in distinguishing between them, and in most places the contact can be approximately mapped from the change in topography. In the view shown in Plate VIII, B, the gentle talus-covered slopes on the left of the canyon above the cliffs of Willow Creek rhyolite consist of Campbell Mountain rhyolite, and so do the similar slopes above the cliffs shown on the right of the canyon in Plate X, A. In the view shown in Plate IX, on the nose between the forks of Willow Creek, the talus-covered, timbered slope between the two cliffs is formed of the Campbell Mountain rhyolite, overlying the lower cliffs of Willow Creek rhyolite and faulted on the north against the upper cliffs of Willow Creek rhyolite. Plate X, B, shows in the left foreground typical outcrops of the Campbell Mountain rhyolite capping the ridge and overlying the cliffs of Willow Creek rhyolite.

DISTINCTION FROM OTHER RHYOLITES.

Locally the Campbell Mountain rhyolite is very difficult to distinguish from the Willow Creek rhyolite; it also resembles some parts of the Windy Gulch rhyolite breccia and the Mammoth Mountain rhyolite. The comparison with the Willow Creek rhyolite will be taken up in the following paragraph; comparison with the Windy Gulch and Mammoth Mountain rhyolites will be left until they have been described.

The normal drab flow breccia of the Campbell Mountain rhyolite is easily distinguished from any part of the Willow Creek rhyolite. Moreover, the normal fluidal, banded Willow Creek rhyolite, as exposed above Creede, is distinct from any part of the Campbell Mountain rhyolite, yet some less typical kinds of what is believed to be the gray rock of the Campbell Mountain rhyolite can not with certainty be distinguished from some of the finely fluidal parts of the Willow Creek rhyolite. In the drainage basins of both forks of Willow Creek, except near Phoenix Park, the two rocks are easily distinguished, and in places at least the contacts are sharp. However, in some places near Phoenix Park as much as 100 feet of rock of doubtful character was found between the typical Campbell Mountain rhyolite and the typical Willow Creek rhyolite, and there appears to be a gradation between them. In the drainage basins of Rat and Miners creeks there is considerably more rock whose position is uncertain, and the boundary between the two rhyolites could not be followed precisely. It is believed that a flow of delicately fluidal rhyolite carrying a few included fragments lies between the normal coarsely fluidal Willow Creek rhyolite and the typical Campbell

Mountain rhyolite. On the map this flow has been included in the Willow Creek rhyolite, which it most resembles. However, poor exposures, alteration, and a variability in the two rocks, which in some places closely resemble each other, have made the mapping unsatisfactory in places. In general the Campbell Mountain rhyolite is a flow breccia with abundant included fragments and a faintly developed flow structure, whereas the Willow Creek rhyolite carries few inclusions and has a prominent and commonly coarse fluidal texture.

PHOENIX PARK QUARTZ LATITE.

GENERAL CHARACTER AND DISTRIBUTION.

The Phoenix Park quartz latite is made up of lava flows and tuff-breccias of a fairly uniform rock; it receives its name from its development about Phoenix Park. It is well developed north of the area shown on Plate II, where it is made up largely of great lava flows, but within the mapped area it consists chiefly of breccia and tuff, with massive flow rock in smaller, less continuous bodies. The breccia is chaotic, shows little bedding or sorting, and consists mainly of rounded or subangular fragments, many of them several feet across. They are nearly all of quartz latite similar to that of the flows. Fine tuffs are only exceptionally present.

This quartz latite was found only in the upper part of the East Willow Creek valley. The main body overlies a fairly regular surface of the Campbell Mountain rhyolite and is overlain irregularly by the Mammoth Mountain rhyolite. Another body of similar rock occurs as a lens between flows of the Campbell Mountain rhyolite. It is believed that here there was an interbedding of the two types of material. This type of rock makes up the greater part of the Alboroto group for many miles about Creede.

This rock wedges out to the south and thickens rapidly to the north. The maximum thickness within the Creede area is about 500 feet, but a few miles to the north it is very much thicker.

PETROGRAPHY.

Megascopic features.—In color the rocks are commonly red-brown to pale reddish drab.[11] They are quartz latites and carry rather abundant visible crystals, about a millimeter in cross section, of white striated plagioclase and glassy orthoclase in about equal amounts, less of glassy quartz, hexagonal plates of black biotite, and prisms of black hornblende and a few of pale-yellow, lustrous titanite. These crystals are embedded in a rather dense to highly porous aphanitic groundmass. For a few feet at the base of some flows the

---

[11] Ridgway's purplish or brownish vinaceous (1′′′b to 5′′′b) to light purple-drab (1′′′′b) or light vinaceous drab (5′′′′b), deepening to livid brown (1′′′) or dark purple-drab (1′′′′i) or becoming as light as pallid vinaceous drab (5′′′′f).

phenocrysts are embedded in a black glass. Many of the rocks carry inclusions of latite of somewhat different color and texture and also of rhyolite, andesite, quartzite, and granular rocks.

In addition to the quartz latites there are a few thin irregular flows of a pale purplish vinaceous [12] rhyolite flow breccia which are characterized by abundant inclusions of fibrous pumice. A few crystals of glassy orthoclase and biotite are present in a rather porous, aphanitic groundmass. This rhyolite is very similar to that characteristic of the Windy Gulch rhyolite breccia.

Microscopic features.—The microscopic study shows that the quartz latites are porphyritic, and that phenocrysts make up nearly half of the rock. These phenocrysts, which range in cross section from 2 millimeters to a fraction of a millimeter and are somewhat broken, include plagioclase, quartz, orthoclase, biotite, hornblende, and titanite, with accessory iron ore, apatite, and zircon. The plagioclase is in well-formed, nearly equant crystals; they are zoned and range from andesine to labradorite, averaging andesine-labradorite. The hornblende is a deep-brown variety in some sections, light green in others, and rather dark olive-green in still others; although commonly nearly as abundant as biotite, it was not found in some of the sections. Titanite occurs in crystals 1 millimeter across and is nearly as abundant as the hornblende. The groundmass is very fine textured and in part spherulitic, in part indistinctly polarizing, and is rhyolitic in character. It is made up mostly of orthoclase but carries a little quartz and probably a very little tridymite in some parts. It is clouded and in addition is dusted with minute inclusions, probably of hematite. The rocks are commonly fresh, but some are altered, probably by hydrothermal agents, and show secondary calcite, sericite, chlorite, and sulphides.

OUTCROPS AND WEATHERING.

The breccia beds are rather easily broken down by weathering, and on the whole the Phoenix Park quartz latite is somewhat less resistant than the overlying and underlying formations. Many of the outcrops are rather poor and are more or less covered with talus or glacial material. North of Phoenix Park flows make up most of the rock and the outcrops are better. To the north of the area mapped, where great flows make up most of this formation, the outcrops are much better, and the very rugged mountain and canyons about San Luis Peak have been carved out of these flows. Weathering is due chiefly to mechanical agencies, and in many places the rock breaks up into a crumbly, sandy mass, although the minerals are still comparatively fresh. Much of the area occupied by this rock has been glaciated.

---

[12] Ridgway's 1‴ *f*.

EQUITY QUARTZ LATITE.

GENERAL CHARACTER AND OCCURRENCE.

The Equity quartz latite, which is made up entirely of massive rock and in large part, at least, of a single great flow, is named from its prominent development near the Equity mine. It overlies rather regularly the Campbell Mountain rhyolite and therefore occupies about the same position in the section as the Phoenix Park quartz latite, from which it differs chiefly in its more massive character but also slightly in its composition, which is somewhat nearer that of a rhyolite. The two latites have not been found in contact, but they are believed to be very closely related and to represent different phases of the same period of eruptive activity.

This rock was found in the upper part of the West Willow Creek basin from upper Deerhorn Creek to and above the Equity mine. It caps the high ridge north of the Equity fault. Just south of the Equity fault it has a thickness of about 1,000 feet, with an erosion surface at the top.

CONTACTS.

The contacts of this flow are nearly everywhere covered with talus, slide, or other débris, and their mapping is not accurate. On the crest of the ridge east of the Equity mine and north of the fault the base is well exposed; here the flow overlies the Campbell Mountain rhyolite rather regularly. Just south of the Equity fault the base is mapped only approximately, as talus and slide completely cover this area. The lowest exposures, which are about 200 feet above the bed of the creek, are the Equity quartz latite; the first exposure in the Equity tunnel is the Campbell Mountain rhyolite; otherwise the mapping of this contact is based on the topography. On the west side of the creek, only a few feet above the creek bed, are good exposures of this latite. North of the Equity mine the trench of the creek is in latite, but the slopes to the east of the flat are in rhyolite, believed to be the Willow Creek rhyolite; there must be a fault between the two. The Western boundary of the mapped body is also believed to be a fault although it is in an area lacking in exposures.

PETROGRAPHY.

Megascopic features.—The fresh rock is commonly light Quaker drab.[13] The hand specimen shows rather abundant crystals of white plagioclase and some of quartz, glassy orthoclase, and biotite in a groundmass which is fairly dense and shows inconspicuous delicate fluidal texture. In addition a few of the rocks show bodies of a pale

---

[13] Ridgway's light Quaker drab (1'''''b) to pale purplish gray (67'''''d), less commonly light olive-gray (23'''''d), light mouse-gray (15'''''b), dawn-gray (35'''''d), or tints or shades of any of these colors.

olive-gray color[14] as much as 5 millimeters wide and a few centimeters long, with a rudely lenticular form but much serrated in detail. They are of coarser crystallization but are not perceptibly porous. The rock is everywhere somewhat altered; the plagioclase and dark minerals are commonly largely replaced by calcite, chlorite, and sulphides. The base of the flow where exposed shows a few feet of dark glass, in which are embedded the usual crystals.

On the upper parts of the ridge included on the map in the Equity quartz latite is a rock which more closely resembles the Phoenix Park quartz latite. In color[15] it is somewhat brighter than the lower part and represents an overlying flow. They are dense fresh rocks with abundant 2-millimeter crystals of white plagioclase and a smaller number of glassy orthoclase, quartz, biotite, green augite, and black hornblende.

Microscopic features.—An examination of thin sections showed that in texture the rock is porphyritic and that the phenocrysts nearly equal the groundmass in amount. These phenocrysts are 3 millimeters in maximum diameter and comprise plagioclase, orthoclase, quartz, and biotite, with accessory zircon, magnetite, titanite, and apatite. The plagioclase crystals range from andesine to sodic labradorite, averaging calcic andesine; they are commonly largely altered to calcite, less commonly to sericite. Some of the larger crystals carry inclusions of small biotite plates and of the accessories. The biotite has been much resorbed by the magma, leaving grains of iron ore along the border; it is also commonly altered to chlorite. Originally there was probably a considerable amount of augite or hornblende present, but it is now completely replaced by calcite and chlorite. Secondary calcite, epidote, and chlorite are rather abundant in all the sections.

The larger part of the groundmass is very finely crystalline to submicroscopic and is rhyolitic in character; it is clouded from submicroscopic inclusions and in addition is dusted with minute brownish trichites. It has a well-developed wavy fluidal texture. Very irregular streaks of the groundmass are made up of a coarser aggregate of quartz and orthoclase in grains or patchy micrographic intergrowths. These areas are irregular in texture and are clear except for scattered trichites of opacite.

The rocks of the upper slopes on both sides of West Willow and Deerhorn creeks differ from those described above in containing more plagioclase and less quartz and orthoclase; the original biotite is almost completely resorbed, grains of augite are fairly abundant, and partly resorbed prisms of brown or green hornblende are present; the groundmass is mainly spherulitic.

---

[14] Ridgway's 23′′′′′*f*.
[15] Ridgway's light purple-drab (1′′′′*b*) to light vinaceous drab (5′′′′*b*), rarely cinnamon-drab (13′′′′).

WEATHERING AND OUTCROPS.

The Equity quartz latite is probably even more resistant to weathering than the Willow Creek rhyolite, but as it occurs only in the upper glaciated parts of the stream valleys, where canyon cutting is not prominent, it forms no cliffs or canyons comparable to those of the Willow Creek rhyolite near Creede.

Near the Equity mine, where a fault separates this quartz latite from the underlying Willow Creek and Campbell Mountain rhyolites, there is a rather striking difference between the outcrops on the two sides of the fault, as shown in Plate VIII, *A*. The Equity quartz latite shows rather rugged outcrops, with jagged, broken cliffs; the rhyolites form steep but smooth talus-covered slopes. Here the Willow Creek rhyolite is less resistant than to the south and does not differ greatly from the Campbell Mountain rhyolite. It yields a scant soil and at the bases of the broken cliffs forms great accumulations of slide and talus. The physical agencies change in temperature and gravity have been most effective in the breaking down of the rock.

Chapter IV.—ROCKS OF THE PIEDRA GROUP.

GENERAL FEATURES.

The rocks which are here included in the Piedra group are separated from both the underlying Alboroto group and the overlying Creede formation and Fisher quartz latite by surfaces of marked irregularity. Both these surfaces are due to erosion during periods of comparatively little igneous activity. The irregular character of these surfaces is not local but has been recognized throughout the eastern part of the San Juan Mountains.

The rocks of this group are chiefly rhyolites and quartz latites, but at about the middle of the group there is a considerable though variable thickness of andesite. The rocks beneath the andesite are chiefly biotite rhyolites and latites; those above are closely related to one another and are hornblende-quartz latites. These hornblende-quartz latites are separated from the underlying flows by a somewhat irregular surface of erosion. Locally the base of the tridymite latite, which is immediately beneath the andesite, is also very irregular. These erosional surfaces do not appear to have been so extensive or so irregular as those at the top and base of the Piedra group.

The rocks underlying the erosional surface at the base of the quartz latites in the western part of the Creede area differ greatly from those occupying the same position in the eastern part, and this difference in the succession of flows and tuff beds for different parts of the area, together with the lack of continuity of the formations through erosion, cover, or lack of deposition, has made the correlations and separations difficult and some of them uncertain. In the western part of the area the lowest formation of this group is a hornblende-quartz latite made up of a chaotic aggregate of thin flows and breccia deposits. It is overlain by the Windy Gulch rhyolite breccia, a series of tuff and flow breccias. This is in turn succeeded by a thick flow of tridymite latite, which is succeeded by the andesite. In the upper part of the West Willow Creek basin are poor exposures of a quartz latite which differs considerably from the tridymite latite but occupies about the same horizon, as it immediately underlies the andesite. In the eastern part of the area the lowest formation of the Piedra group is a great flow of the Mammoth Mountain rhyolite, a flow breccia. It is overlain by a rhyolite tuff with associated thin flows of rhyolite, and these are in turn overlain by the tridymite latite. The andesite is absent.

To the east of West Willow Creek above the mouth of Deerhorn Creek the section is still somewhat different, but exposures are too

35

poor for a positive determination of the relations existing between the rocks of this area. The tridymite latite appears to be the lowest rock exposed. It is overlain by 50 feet or so of an andesite which is believed to be the same as the andesite that caps Bulldog Mountain. This is overlain by about 100 feet of thin-bedded rhyolite tuff, which probably corresponds to the tuff that overlies the andesite to the west. This in turn is succeeded, at least on the slopes, by a rhyolite which has not been recognized elsewhere in the area and which is an unusually thick flow in the quartz latite tuff. Several hundred feet of chaotic tuff-breccia overlies this flow.

The quartz latites that overlie the andesites, or, in their absence, the tridymite latite or one of the older formations, are very similar to the Phoenix Park quartz latite. The lowest formation of these, the quartz latite tuff, is made up mostly of tuff but has some thin flows. Over this tuff is the Rat Creek quartz latite, which is made up mostly of flows with some tuff; the upper formation is a mesa-forming flow of quartz latite about 200 feet thick, the Nelson Mountain quartz latite.

The thickness of the Piedra group, as might be expected from a series of igneous rocks with irregular surfaces at both the top and bottom and within it, varies greatly from place to place. Within the Creede area it is probably greatest east of Rat Creek, where it is over 2,000 feet.

HORNBLENDE-QUARTZ LATITE.

GENERAL CHARACTER AND DISTRIBUTION.

In the western part of the quadrangle the lowest mapped formation included in the Piedra group is a rather chaotic aggregate of small irregular flows and clastic material. The rocks vary considerably in character but are chiefly quartz latites with conspicuous hornblende, and the whole mass is therefore called a hornblende-quartz latite, although hornblende andesite is common and rhyolite is present. This rock occurs on the small hill just northeast of Monon Hill. A more extensive and continuous but thin layer extends westward nearly on the contours from a point southwest of Bulldog Mountain to the fault near the Kreutzer mine. West of the fault it is nearly 1,000 feet higher on the slopes and continues westward into the basin of Shallow Creek but has not been found farther west and appears to be a very local body.

THICKNESS.

As this latite overlies the rocks of the Alboroto group irregularly, and as its top is fairly regular, its thickness varies considerably. Within the Creede area it is probably nowhere over 200 feet, but higher up Miners Creek and on Shallow Creek it is considerably thicker.

PETROGRAPHY.

Megascopic features.—In color the rocks vary considerably; the greater part are drabs.[16] Most of the fresh rocks are rather dense, and nearly all show a more or less prominent fluidal structure. Hornblende, which is the most abundant dark mineral, is conspicuous from its lustrous black cleavage faces; biotite in the usual black flexible flakes is commonly almost as abundant; and green pyroxene can rarely be seen with a pocket lens. White plagioclase crystals are abundant, and in some of the rocks many of the crystals are several millimeters across. Colorless orthoclase is considerably less abundant and in many of the rocks is lacking; large dull orthoclase crystals several centimeters across are sparsely present in some of the rocks; quartz in visible crystals is rare. The groundmass, which is about equal in amount to the phenocrysts and is holocrystalline, can not be resolved with a pocket lens, but in most of the rocks it has the appearance of a rhyolite; in a few it looks like an andesite. At the bases of some of the flows there is a few feet of black obsidian carrying the usual phenocrysts. On the whole the rocks are not very different from some of the quartz latites of the Alboroto group, especially the Outlet Tunnel quartz latite.

Microscopic features.—The study of thin sections showed that the greater part of the rocks are quartz latites. The hornblende is brown or less commonly green; both varieties are present in distinct crystals in some thin sections. It has usually been considerably resorbed by the magma; the biotite has been resorbed to a less extent. The plagioclase is andesine or andesine-labradorite with a more sodic border. The accessory minerals are pleochroic apatite, magnetite, and rare titanite. The groundmass varies considerably; in some specimens it is spherulitic, in others microfelsitic, microgranular, micropegmatitic, or in part glassy; it is made up chiefly of quartz and alkalic feldspar. Secondary chlorite, calcite, and kaolinite are present.

The andesites, which are not abundant, differ from the quartz latites chiefly in the character of their groundmass, which is made up largely of minute laths of oligoclase feldspar with some rods and grains of pyroxene. The phenocrysts are somewhat more abundant, and in a few of the rocks those of feldspar are tabular. The rocks as a whole are near the border between the quartz latites and the andesites; those called andesites do not differ greatly from the quartz latites, and there is a fairly continuous gradation between the two. In this they differ considerably from most of the other quartz latites of the area, which are considerably nearer the rhyolite.

---

[16] Near Ridgway's pale Quaker drab (1′′′′′*d*) or purple-drab (1′′′′′); less commonly olive-buff (21′′′*d*) or a related color.

Locally the rocks are greatly altered by processes other than weathering. The plagioclase is kaolinized, the hornblende is altered to chlorite, iron oxide, and other minerals, and the rock is bleached to a dirty light gray or white.

WEATHERING AND OUTCROPS.

The breccia portion of this formation breaks up readily and offers few outcrops; the massive rock is somewhat more resistant, but it too is a comparatively soft rock and rarely gives good outcrops. Within the Creede area this quartz latite is thin and has not had a great influence on the development of the topography.

WINDY GULCH RHYOLITE BRECCIA.

GENERAL CHARACTER AND DISTRIBUTION.

A rhyolite breccia made up of light-colored rhyolite, in part an ordinary tuff, in part a normal flow rock, but chiefly a breccia, probably a flow breccia, lies beneath the tridymite latite in the drainage basin of Windy Gulch and to the west and is here called the Windy Gulch rhyolite breccia. Its porous character and the abundant fragments of pumice which it carries are characteristic.

It overlies a fairly regular surface of the hornblende-quartz latite in the drainage basins of Rat and Miners creeks, but to the east it directly overlies the Campbell Mountain rhyolite. Near its base it commonly carries very abundant inclusions of the underlying rock. It is everywhere overlain rather irregularly by the tridymite latite. Southeast of Bulldog Mountain it is in contact with the Creede formation, which lies below it on the slopes. It forms a layer, broken in places by faulting or by a Quaternary covering, from a point west of MacKenzie Mountain to 'the Happy Thought mine. Exposures are very poor, and the lower contact in particular can in few places be mapped with accuracy. The body west of MacKenzie Mountain is poorly exposed, and its lower contact especially is uncertain. On the east side of the Kreutzer fault the rhyolite breccia is thrown down nearly 1,000 feet; the apparent thickness here is probably due to a disturbance of the beds near the fault, but exposures are almost lacking. East of the fault the rock continues to Rat Creek and is nearly horizontal; beyond Rat Creek it rises rapidly and crosses Windy Gulch just below Bachelor. Beyond Windy Gulch this rhyolite breccia, aside from a small outcrop just south of Bachelor, is mapped only as a narrow band west of the Amethyst fault and between the Last Chance and Happy Thought mines. This is in an area that is almost completely covered with a mantle of débris from the slopes above and probably also of ancient wash and is almost entirely lacking in exposures. Not many of the numerous prospect shafts can be entered, but their dumps furnish sufficient data, although the rocks are commonly much altered, to map this area with a fair degree of accuracy. The breccia is probably local in extent, as outside the area mapped on Plate II it has been recognized only to the west, as far as Shallow Creek.

This rhyolite is commonly from 100 to 200 feet thick, although locally it is probably much thicker, as it is believed to have spread over a surface of some relief. However, it has been faulted and considerably disturbed east of Miners Creek and south of Bulldog Mountain, and the greater apparent thickness in both these areas is due, at least in part, to local tilting associated with faulting.

PETROGRAPHY.

Megascopic features.—In color this rock is commonly pale red-brown;[17] less commonly it has one of the lighter tints of gray; the altered rock is white or gray. The luster is always chalky dull. The rock is made up in considerable part of irregular, ragged fragments, rarely over a few centimeters across, of pumice having a fibrous appearance, due to the fine elongated pores. These fragments are of somewhat lighter color than the body of the rock; locally they are replaced by a green claylike material or by drusy quartz. They weather out, leaving very numerous rough cavities of various sizes. In addition there are fewer fragments, commonly not over 1 centimeter across but rarely over 1 decimeter, of a variety of foreign rocks; these are generally much altered and consist chiefly of the Alboroto rhyolites and the hornblende-quartz latite, but a few are other rhyolitic and andesitic rocks. The matrix is always less porous than the pumice fragments, and in some of the rock it is fairly dense. It carries a few 1-millimeter crystals of glassy orthoclase and a very few of biotite. Much of the rock is believed to be a flow breccia, although a part may represent a "mud flow" or tuff. The fragments, which make up a large part of the rock and are similar to the matrix except for their greater porosity, probably represent fragments that were shot into the air and fell back into the molten magma. These flow breccias may also represent material that was thrown from a crater, settled about the vent while still soft, became welded together, and flowed much as any viscous lava.

Microscopic features.—The thin sections of the rock show a few crystals of orthoclase and a very few of plagioclase, biotite, magnetite, apatite, and zircon, in a highly porous, glassy groundmass.

---

[17] Ridgway's purplish vinaceous (1'''b) to pale purplish vinaceous (1'''f), brownish vinaceous (5'''b) to pale brownish vinaceous (5'''f), pale vinaceous (1''f), pale purple-drab (1''''d), or some closely related color.

WEATHERING AND OUTCROPS.

The Windy Gulch rhyolite breccia is considerably softer than the overlying tridymite latite or even than the Campbell Mountain rhyolite, which underlies it in places. It forms smooth, grass-covered or timbered slopes, with rock outcrops only in very favorable places. On exposure to weathering the pumice fragments are removed and the rock breaks up into honeycombed fragments with considerable interstitial fine material. Where erosion is not too rapid a soil is formed in which are few rock fragments and these chiefly of the latitic inclusions.

COMPARISON WITH CAMPBELL MOUNTAIN RHYOLITE.

The typical rocks of the Campbell Mountain rhyolite and of the Windy Gulch rhyolite breccia can readily be distinguished. They differ considerably in color, and much of the breccia can be recognized by its porosity and especially by the abundant pumice fragments which it includes. However, the denser, darker-colored varieties of the Windy Gulch rhyolite breccia are almost identical with some of the lighter-colored, less dense varieties of the Campbell Mountain rhyolite. On weathering both lose their inclusions, become highly cavernous, and are still more difficult to distinguish. Areas that caused especial difficulty in the mapping are on the ridge north of the Corsair mine, on the ridge south of Bulldog Mountain, and northeast of Monon Hill.

MAMMOTH MOUNTAIN RHYOLITE.

GENERAL CHARACTER AND DISTRIBUTION.

The Mammoth Mountain rhyolite is a single thick flow of rather uniform character. It is confined to the northeastern part of the Creede area and to the mountains to the east and northeast. The small triangular body in the upper part of Dry Gulch, east of Mammoth Mountain, is poorly exposed; it is probably bounded by faults. The greatest body is north of this area and directly overlies the rocks of the Alboroto group. A considerable body lies on the ridge between the forks of Willow Creek. This flow has not been recognized to the west, south, or north of the Creede area, but it has been followed to the east as far as Bellows Creek.

CHARACTER OF CONTACTS AND THICKNESS.

The base of this flow is very irregular, as may be seen from an examination of Plate II. The irregularity is shown east of East Willow Creek, where the lower contact commonly cuts sharply across the contours, and several of these contacts were at first thought to be faults. The flow generally overlies the Phoenix Park quartz latite or the Campbell Mountain rhyolite but locally rests on the Willow Creek rhyolite. It occupies about the same position in the section as the hornblende-quartz latite and the Windy Gulch rhyolite breccia, to the west, but its relation to these rocks could not be determined. Overlying the Mammoth Mountain rhyolite fairly regularly is the rhyolite tuff or the quartz latite tuff.

East of East Willow Creek this flow attains a thickness of 1,000 feet, but the thickness diminishes rapidly to the west, and the flow wedges out in the upper drainage basin of Nelson Creek. It is believed that the lava came from the northeast or east and did not extend far west of Campbell Mountain.

PETROGRAPHY.

Megascopic features.—In color this rock is characteristically red-brown.[18] Its luster is dull; its fracture is commonly rough and tends to be hackly. It is a flow breccia and is decidedly mottled, though somewhat less strikingly so than the Campbell Mountain rhyolite, which it very closely resembles. It shows a fluidal texture only on close examination, with very fine wavy streaks, discontinuous and irregular. The rock is fairly dense, although cavities a millimeter or so across are sparingly present. Much larger cavities, in part flattened, are exceptional but are rather abundant locally; they are probably confined to the base and top of the flow. They are in part filled with kaolinitic material. At the base of the flow wherever seen there is a few feet of black glass.

The rock contains rather abundant phenocrysts as large as 2 millimeters in cross section; they make up approximately 10 per cent of the rock. Glassy orthoclase is in slight excess over porcelain-white plagioclase; quartz and biotite are less abundant. The groundmass is aphanitic. The mottled appearance is due to the presence of abundant inclusions, which are of two kinds—one distinctly foreign, the other differing from the host chiefly in having a slightly paler color. The former are commonly much altered and are chiefly quartz latites similar to those of the underlying Alboroto group; less common ones are rhyolites similar to those of the Alboroto group or more coarsely porphyritic rhyolites or andesites. Some of them are bounded by narrow bands of lighter or darker color.

Microscopic features.—The thin sections show that the plagioclase of the phenocrysts is oligoclase or andesine; zonal growths are inconspicuous. The biotite is slightly resorbed. Minute to submicroscopic particles of ferritic material give the groundmass a pinkish-buff color [19] or a paler tint as seen in a thin section of normal thick-

---

[18] Ridgway's russet-vinaceous (9′′′); rarely a darker shade and very rarely a lighter tint; exceptional specimens contain less gray; more common ones contain more gray, as brownish drab (9′′′′), or more orange, as fawn-color (13′′′).

[19] Ridgway's 17′′*d*.

ness. The groundmass is not markedly fluidal but shows irregular, discontinuous wormlike streaks of slightly different color and texture. It is largely submicroscopic in crystallization but in part is delicately spherulitic, with the minute fibers arranged normal to the fluidal streaks. It has the characteristics of a rhyolite groundmass and is no doubt made up chiefly of quartz and orthoclase with some tridymite. Besides the streaks mentioned above and others much larger, some several millimeters across, the sections show irregular lenses and streaks of colorless material. These are much more coarsely crystalline than the body of the groundmass and are made up of microscopic intergrowths of quartz and orthoclase; commonly they show a concentration of quartz in their centers. Small rounded areas, rarely a millimeter across, are abundant in some of the sections and are largely confined to the coarsely crystalline areas. They are made up of a mineral which has an index of refraction of about 1.50 and a very low birefringence and which is probably a zeolite. Some of the original cavities are filled with a fibrous or platy kaolinitic material which has an index of refraction of 1.535 and a moderate birefringence; in some specimens this is deposited on the spherulites mentioned above.

In addition to the inclusions of quartz latites there are common inclusions of a rhyolite porphyry with phenocrysts of orthoclase, plagioclase, quartz, and sphene in a fairly coarse granophyric groundmass. Fragments of altered andesite are rare. Other inclusions of rhyolite differing slightly from the host but sharply bounded and commonly having a more coarsely crystalline groundmass and paler tint are abundant. Many of them are clearly included fragments, but some can hardly be distinguished from the more coarsely crystallized portions which are believed to have crystallized in their present position.

WEATHERING AND OUTCROPS.

This rhyolite is a rather resistant rock and has weathered largely through mechanical agencies. It has no banding or regular direction of weakness, but the weathering is influenced by the inclusions and the irregular, streaked structure, so that the rock disintegrates into very irregular, hackly fragments commonly about the size of a pea; on the steeper slopes, where the loose material is carried away as rapidly as it is formed, the outcropping rock is very rough and hackly. The rock is uniform in character, structureless for 1,000 feet in thickness, and more resistant than either the underlying quartz latite or the overlying tuff. It gives rather prominent outcrops that are rough in detail and steep, fairly regular slopes with no prominent cliffs or benches. It is best exposed along First Fork of East Willow Creek.

COMPARISON WITH CAMPBELL MOUNTAIN RHYOLITE.

The Mammoth Mountain rhyolite is easily distinguished from all the associated rocks except the Campbell Mountain rhyolite, which it very closely resembles. Both rocks are flow breccias and contain foreign inclusions of quartz latite and other rocks, in addition to those differing from the host chiefly in their paler tint; both rocks are rather dense, with few small gas cavities; both have a dull luster; both show indistinct and poorly developed flow structure. On the whole they differ slightly in color, but the color variation in either one is greater than the difference between the two. The Campbell Mountain rhyolite is commonly a little duller or more brownish than the Mammoth Mountain rhyolite.[20] The phenocrysts are about the same in both rocks. The microscopic examination aided but little in their distinction. In brief, the difference between specimens of the Campbell Mountain rhyolite, especially between the drab and gray types, is much greater than that between the drab type of the Campbell Mountain rhyolite and the typical Mammoth Mountain rhyolite, and if a number of typical hand specimens of each were mixed together they could not be separated except by one who was thoroughly familiar with both rocks. Only after a careful study of both types and of numerous specimens from each has the writer been able to distinguish between the two with reasonable assurance.

One of the most constant and characteristic differences between the two rocks is in the weathering. The Campbell Mountain rhyolite weathers into flat flakes with smoothly rounded surfaces that may be a foot or so across; it breaks with a smooth conchoidal fracture. The Mammoth Mountain rhyolite almost invariably weathers into small, irregular hackly fragments, most of them less than an inch across; its outcrops nearly everywhere show the character of the weathering, and specimens broken from even apparently fresh rock show an irregular, hackly fracture; only exceptionally do they show a smooth conchoidal fracture. The difference in color, as stated above, is not in itself conclusive but aids in distinguishing between the two rocks. The general appearance and inclusions of the two rocks are somewhat different. In the Mammoth Mountain rhyolite the fragments of rhyolite similar to the host except in color are less conspicuous, hence the rock is less prominently mottled than the Campbell Mountain rhyolite. Inclusions of nearly white porphyritic rhyolite are locally characteristic of the Mammoth Mountain rock. The phenocrysts, especially those of biotite and plagioclase, are

---

[20] According to Ridgway's color nomenclature the Campbell Mountain rhyolite is commonly purple-drab (1''''') or a nearly related color, and the Mammoth Mountain rhyolite is commonly near russet-vinaceous (9''').

somewhat more abundant in the Mammoth Mountain rhyolite, and the whole groundmass has a wavy, chaotic delicate flow banding. In the thin sections a chaotic, discontinuous wormlike fluidal banding is rather characteristic of some of the rocks of the Piedra group. The black glass at the base of the Mammoth Mountain rhyolite and the cavernous rock which is locally present in it aid in separating it from the underlying rhyolite. The presence between the two rocks of a small amount of Phoenix Park quartz latite, even though not in sufficiently definite bodies for mapping, has aided in separating the two.

After a careful second examination of much of the contact west of East Willow Creek the separation has been made with more confidence and accuracy than was at first believed possible. The greatest difficulty was experienced in the area just west of Phoenix Park, where exposures are poor and much of the rock is of doubtful character. The final mapping includes in the Mammoth Mountain rhyolite some outcrops which occur just south of the small stream that passes through Phoenix Park and which from the specimen alone might be included in the Campbell Mountain rhyolite, although they are not typical.

RHYOLITE TUFF.

GENERAL CHARACTER AND DISTRIBUTION.

On the upper slopes to the east of Phoenix Park and in large part beyond the boundary of the area mapped a siliceous tuff with associated thin flows of rhyolite rather regularly overlies the Mammoth Mountain rhyolite and is in turn overlain regularly by the tridymite latite. It is about 200 feet thick and has one or locally two thin flows near its base. The main part is a sandy tuff which in places carries scattered pebbles; it is very poorly sorted and bedded.

PETROGRAPHY.

The tuff is nearly white to pale drab and has in large part a sandy texture. It carries very abundant glassy crystals of feldspar and black biotite several millimeters across and a few small fragments of a rock which contains abundant phenocrysts in a pumiceous matrix. Larger pebbles of similar rock and of a variety of other rocks are locally present.

A part of the material is a quartz latite tuff made up largely of broken crystals of plagioclase with considerable orthoclase, some quartz, biotite, green hornblende, and augite in a very fine glassy matrix and closely resembles the quartz latite tuff under Nelson Mountain.

The flows are rarely over 25 feet thick, and where two are present they are separated by a small amount of tuff. The rock of the flows is a rather porous light red-brown [21] rhyolite flow breccia with numerous small fragments of pumice and rhyolite and a few of a fine-textured granitic rock. It carries scattered glassy crystals of orthoclase, white andesine, and black biotite with accessory apatite, zircon, and magnetite. The groundmass is largely glass, with numerous trichites of red hematite and some ropy streaks that are largely crystalline.

TRIDYMITE LATITE.

GENERAL CHARACTER AND OCCURRENCE.

The tridymite latite is the most nearly uniform and distinctly characterized rock of the Piedra group. The main flow throughout the area studied is a banded fluidal rock with lenses or layers which are highly porous and which are filled with minute drusy crystals of tridymite. Near Bachelor and elsewhere a denser rock free from tridymite forms the base of this formation and probably represents a different flow. The tridymite latite is confined to the area west of the Amethyst fault. The main body forms a band from the west slopes of MacKenzie Mountain to the Happy Thought mine; it is commonly almost horizontal but is displaced by several faults and locally near the faults dips steeply. Small isolated bodies of the typical rock are present about half a mile west of Sunnyside and others north of Monon Hill; isolated bodies of less typical rock are present on upper Rat Creek and northwest of the mouth of Deerhorn Creek. The presence of the bodies north of Monon Hill and west of Sunnyside can be explained only on the assumption that the tridymite latite flowed over a surface of greater relief in this area than has been recognized elsewhere. The correlation of these bodies is based entirely on the character of the rocks, without confirmation from the sequence of flows. Such a correlation is rarely beyond question. In the Creede area, however, the tridymite latite is a unique rock and has been found at only one horizon.

The details of distribution within the Creede area are shown on Plate II. To the west of the area included on the map the upper surface of the tridymite latite forms the great flat north of Bristol Head; to the east it underlies the great bench called Wason Park. A similar rock, which is probably a part of the same flow or a very closely related flow, is present south of the Rio Grande in the drainage basin of Trout Creek. Nearly everywhere it immediately underlies the latite tuff, but under Bristol Head the two are separated by andesite.

It is not present on the slopes of Nelson Mountain, although it covers great areas both to the east and west. It was probably

---

[21] Ridgway's light russet-vinaceous (9'''b).

locally eroded from this area preceding the deposition of the quartz latite tuff although it may never have covered the area where the Nelson Mountain ridge now stands.

THICKNESS.

The thickness of this latite varies greatly and rapidly from place to place, due chiefly to the irregularity at its base. Its greatest thickness is about 400 feet beneath Bulldog Mountain; from this it decreases to 50 feet west of MacKenzie Mountain.

PETROGRAPHY.

Megascopic features.—In color the rock is rather dark red-brown.[22] It is characteristically banded and platy; the main part is rather dense, but irregular streaks and lenses of a paler tint, as much as 1 centimeter across, are decidedly porous. These are rather evenly spaced a few centimeters apart and make up a considerable part of the rock. The rock shows phenocrysts of white plagioclase, glassy orthoclase, and black biotite nearly equal in amount to the groundmass as much as 2 millimeters across. The pores are characteristically lined with minute drusy crystals of tridymite.

Microscopic features.—The microscopic examination showed that the phenocrysts consist of orthoclase and plagioclase in nearly equal amount, considerable biotite, and accessory apatite, magnetite, and zircon. The plagioclase crystals have a core of andesine or andesine-labradorite grading through two or more rather broad intermediate zones to a narrow border of albite or oligoclase; their average composition is about that of andesine. They are commonly in part altered to sericite and kaolinite. The biotite has been more or less resorbed by the magma, with the separation of iron oxide. The groundmass is beautifully fluidal, with wavy bands. The main part is so finely crystalline as to be only indistinctly polarizing; it is clouded and reddish brown in reflected light from numerous microscopic specks of hematite. It is probably made up largely of orthoclase but may have some tridymite. There are numerous lenses or streaks of clear material which has a much coarser crystallization, and the larger ones are porous. They range in width from a fraction of a millimeter to several millimeters. They vary greatly in coarseness of crystallization and are in part spherulitic or fibrous, in part micrographic or microgranular. Orthoclase and tridymite are the chief minerals of these streaks. Tridymite, which is very abundant, is present as aggregates filling rounded areas or lining the walls of the

---

[22] Ridgway's vinaceous brown (5'''*i*) or a nearly related color; the paler tints, deep brownish vinaceous (5''') and brownish vinaceous (5'''*b*) are common, as are also the colors with more red, livid brown (1''') and purplish vinaceous (1'''*b*), and those with more gray, purple-drab (1'''') and dark purple-drab (1''''*i*); paler tints and darker shades are exceptional.

cavities and is evidently closely associated with the gas cavities. In a few specimens tridymite is absent, and its place in the centers of the coarsely crystalline bands is taken by quartz. In one specimen from upper Rat Creek the cavities are lined with botryoidal opal.

The origin of these coarsely crystalline porous bands is believed to have been much the same as that of the somewhat similar bands in the Willow Creek rhyolite and has been discussed on pages 20–23, in connection with the description of that rock. However, the conditions were not identical, as in this flow a large part of the silica is in the form of tridymite and the amount of space occupied by the gas cavities is considerably greater.

TRIDYMITE, QUARTZ, AND CRISTOBALITE IN VOLCANIC ROCKS.

The most characteristic feature of this rock is the abundance of tridymite it contains. Some of the other rocks of the Creede district also carry considerable tridymite, and, indeed, in the whole San Juan region of Colorado tridymite occurs in many of the rhyolites and latites, and in a considerable number it is one of the chief constituents, forming from one-fifth to one-third of the rock. In some rocks both quartz and tridymite are present, quartz usually as intratelluric phenocrysts and tridymite in the groundmass, as one of the latest constituents to crystallize. Tridymite is commonly either closely associated with the gas pores or with spherulitic and similar growths. Tridymite is also fairly abundant in the andesitic and basaltic rocks, where it is confined to the gas cavities and is commonly associated with albite and pale-brown biotite and less commonly with needles of hornblende. In the San Juan region tridymite is present in nearly as large a proportion of the volcanic rocks as quartz, and it is also present in nearly as large amount and is therefore one of the chief rock-forming minerals of the region. From the writer's observations on numerous volcanic rocks in other parts of the western United States, tridymite appears to be a much more common and abundant constituent in volcanic rocks than is generally recognized.

Cristobalite has also been observed in many of the rocks of the San Juan region and other parts of the western United States, though it is present in fewer rocks and is less abundant than tridymite. It has been found in rhyolite and quartz latite but appears to be as common in andesite and basalt. Wherever identified it is present in small amounts in rounded spherulitic growths, which grade by orientation of the spherulites into poorly formed crystals with complex twinning. These rounded grains are perched on the walls of gas cavities.

Of the three forms of silica, only quartz has been formed as intratelluric phenocrysts. Tridymite is confined to the groundmass and especially to the more porous, coarsely crystalline parts of the groundmass that were richer in mineralizers and that crystallized last. Quartz is also found in this same association, and the conditions that govern the formation of quartz or tridymite have not been determined, although a particular flow or group of related flows usually carries either one or the other over a large area. Thus in the Creede area the Willow Creek rhyolite carries quartz, and the tridymite latite carries tridymite. In some flows both minerals are present, apparently in close association. The alteration of one to the other has nowhere been observed. However, it is a common phenomenon for quartz phenocrysts to be largely resorbed and tridymite to be later deposited in the gas cavities or groundmass. As cristobalite has been found only as scattered crystals perched on the walls of the larger gas cavities and as in some andesitic rocks it appears to be associated with alteration of the rock, it is probably formed at a later stage in the crystallization, after complete solidification of the main part of the rock. In some specimens, especially of andesitic and basaltic rocks, tridymite also is confined to the vesicles and has much the same relation to the normal groundmass as cristobalite, although it is commonly much more abundant, may nearly fill the vesicles, and is more closely associated with albite, biotite, and hornblende. On the whole, the writer's observations indicate that in volcanic rocks quartz forms in the magma chamber and throughout the crystallization of the groundmass, tridymite in large part during the later stages of the crystallization of the groundmass and after the solidification of the main groundmass, and cristobalite at a somewhat later stage. All three commonly form in the presence of abundant mineralizers, but quartz may crystallize as phenocrysts and in dense rocks in the absence of any unusual proportion of mineralizers, whereas tridymite is nearly always accompanied by abundant mineralizers, as is shown by its close association with biotite and gas cavities, and cristobalite appears to require an even greater abundance of mineralizers, as is indicated by its sparse occurrence in relatively large gas cavities.

It is not intended in this place to discuss in detail the chemistry of the deposition of silica in volcanic rocks. However, the common association of tridymite with gas cavities and biotite shows its close association with the volatile constituents, and its association with only small amounts of other minerals, together with the fact that it commonly occupies about as large a volume as the gas cavities, shows that it could hardly have been deposited from an ordinary solution of silica in the water and other mineralizers. It seems far more likely that in the final fluid part of the magma the silica and associated mineralizers were in chemical combination, and that on the breaking up of these compounds the tridymite was deposited and the volatile constituents released, thus forming the gas cavities. These compounds may have given the rhyolitic magmas their remarkable fluidity and have prevented the rapid escape of the volatile constituents, as the lavas spread in thin sheets over large areas. The conditions do not seem to be very different for much of the quartz in the groundmass, though some of it is less closely associated with gas cavities and more closely associated with feldspar. It may easily be that the composition of the so-called mineralizers largely determines the form in which the silica is deposited.

CHEMICAL COMPOSITION.

A specimen obtained about 100 yards northwest of the schoolhouse at Bachelor was selected for analysis as representing the typical tridymite-rich rock. The material was taken across the bands, so as to represent the average of the flow. The plagioclase is slightly kaolinized, but otherwise the rock is fresh. An analysis was made by George Steiger in the Geological Survey laboratory and is as follows:

Analysis of tridymite latite.

SiO_2	67.76	TiO_2	0.45
Al_2O_3	16.08	ZrO_2	.02
Fe_2O_3	2.22	CO_2	None.
FeO	.23	P_2O_5	.11
MgO	.43	S	.02
CaO	2.59	MnO	.04
Na_2O	4.06	BaO	.12
K_2O	4.91	SrO	.03
H_2O-	.94		
H_2O+	.54		100.55

The norm computed according to the quantitative classification is as follows:

Norm of tridymite latite.

Quartz	19.78	Ilmenite	0.61
Orthoclase	28.91	Titanite	.39
Albite	34.06	Apatite	.34
Anorthite	11.40	H_2O, etc	1.48
Hypersthene	1.00		
Diopside	.22		100.41
Hematite	2.22		

The mode differs from the norm chiefly in the presence in the former of tridymite instead of quartz and of biotite instead of pyroxene.

The rock is a toscanose (I.4.2.3).

FLOW NEAR MOUTH OF DEERHORN CREEK.

On both sides of West Willow Creek just above the mouth of Deerhorn Creek there are a few small exposures of a quartz latite that differ somewhat from the normal tridymite latite. Similar material was found on the dump of the Captive Inca shaft.   Its base is nowhere exposed, and the overlying rocks are but poorly exposed on the northeast side of the creek.   It is immediately overlain by an andesite resembling the andesite of Bulldog Mountain and is believed to be closely related to the tridymite latite and has been mapped as belonging to that body.

In color the rock is near Quaker drab.[23]   The fresh rock shows a rather delicate, inconspicuous, wavy fluidal banding which is more conspicuous on the somewhat altered rock.   The rock is rather dense and shows scattered crystals, approximately 3 millimeters in cross section, of white plagioclase and clear, glassy orthoclase and a very few of biotite.   In much of the rock the plagioclase is altered to kaolinite and sericite.

The thin sections show that the rock contains phenocrysts of orthoclase and plagioclase in about equal amounts with less of biotite and accessory minute crystals of apatite, zircon, and magnetite.   The groundmass is fluidal and is in part spherulitic, in part microfelsitic. The rock differs from the typical tridymite latite chiefly in that it lacks the prominent tridymite-rich lenses.

OUTCROPS, WEATHERING, AND ALTERATION.

The tridymite latite is considerably more resistant than the Windy Gulch rhyolite breccia that commonly underlies it and is somewhat more so than the overlying andesite.   It is commonly traversed by a system of vertical joints, and these, together with the horizontal fluidal structure, determine its mode of weathering.   It breaks into small plates ôr irregular fragments, with little chemical decomposition.   It is a cliff-forming member, and as it is underlain by a comparatively weak rock several large landslides, notably the one west of Bulldog Mountain, have broken from its cliffs.

The mesa and prominent bordering cliff against the snow-covered mountain in the background shown on the right of Plate IV, B (p. 3), are formed by this flow.

It is bleached and silicified near the Amethyst vein, although the orthoclase crystals are still fresh.   Nearly everywhere the plagioclase crystals show some alteration to kaolinite or less commonly to sericite, and locally they are completely gone, although the remainder of the rock shows no alteration.   This alteration probably took place after the crystallization but before the cooling of the magma and was probably caused by the gases that occupied the pores in the lava.

---

[23] Ridgway's Quaker drab (1′′′′′′), light Quaker drab (1′′′′′′b), or light purple-drab (1′′′′b).

ANDESITE.

GENERAL CHARACTER AND DISTRIBUTION.

Andesites and related rocks occupy only a very small part of the Creede area. The only large mass of rock of this character overlies the tridymite latite and underlies the quartz latite tuff. This mass is made up of a considerable number of thin flows with a somewhat smaller amount of intercalated breccia. The rocks vary considerably and include biotite-augite-quartz latites, biotite-hornblende andesites, augite andesites, and olivine andesites. Normal basalts have not been recognized.

The largest area occupied by these rocks is on Bulldog Mountain and the ridge to the north. In the upper Rat Creek basin are two small outcrops, and west of Rat Creek a narrow band of this andesite occurs above the tridymite latite. Along West Willow Creek just above and below the mouth of Deerhorn Creek there are two small outcrops of massive rock which are believed to belong to this mass. Farther east this rock has not been found. West of the Creede area, however, on Bristol Head, these andesites are locally well developed and occupy their usual position, regularly overlying the tridymite latite and rather irregularly overlain by the quartz latite tuff.

Throughout the area over which this andesite has been studied it is irregular in thickness and discontinuous; the base is fairly regular, and the greater part of the variation in thickness is due to the irregularity at the top. The greatest thickness is shown north of Bulldog Mountain, where there is about 500 feet of this andesite, but it becomes rapidly thinner in all directions.

PETROGRAPHY.

General features.—In color the rocks are largely near Quaker drab or mouse-gray.[24] The greater part of the rocks are fine textured and carry very few crystals that can be seen without a careful inspection with a pocket lens. They are as a rule conspicuously vesicular or amygdaloidal. In many places the vesicles are much flattened by flow, giving the rocks a platy, fluidal texture. Megascopically they were thought to be basalts. The lowest flow in the drainage basin of Windy Gulch is dark reddish brown[25] and rather dense. It carries scattered crystals of white plagioclase and brownish-black biotite and hornblende a few millimeters across in a felsitic groundmass. On the east bank of West Willow Creek just above the first crossing of the road to the Equity mine is a small exposure of a dense deep mouse-gray[26] rock which carries abundant tabular

---

[24] Ridgway's Quaker drab (1'''''') or deep mouse-gray (15'''''i) or a tint or shade of either of these; rarely purple-drab (1'''') or olive-gray (23'''''b).

[25] Ridgway's dark purple-drab (1''''i).

[26] Ridgway's 15'''''i.

crystals of glassy, striated plagioclase as much as 5 millimeters. long and a little augite in an aphanitic groundmass. A similar rock was found under the white tuff in a shallow shaft a few hundred feet to the south, where it is associated with the normal amygdaloidal rock. Rock of this type was also found at the bottom of a shallow shaft west of West Willow Creek and S. 25° W. of the mouth of Deerhorn Creek, at an elevation of 10,900 feet.

Microscopic features.—As seen under the microscope the typical amygdaloidal rock contains rather abundant 1-millimeter tablets of plagioclase, with a little augite and altered olivine, which grade into a very fine-textured groundmass made up of plagioclase laths and a somewhat smaller amount of augite grains and magnetite, with considerable brown, clouded interstitial glass. A small amount of quartz is present in some of the rocks; it may be secondary, although it appears to be interstitial to the feldspars of the groundmass. The larger feldspar crystals have cores of labradorite or sodic labradorite and grade outward into albite. The outer zones are filled with glass inclusions. The average composition of the phenocrysts is near andesine. The feldspars of the groundmass range from albite to andesine. The olivine, which is variable in amount, is completely altered to iddingsite, fibrous strongly birefracting serpentine, carbonates, and iron oxide. The augite is commonly reddish brown from the partial oxidation of the iron, probably before the consolidation of the rock. Some of the sections show areas of opacite, which probably represent resorbed crystals of amphibole or other dark minerals. A few of the rocks are holocrystalline and carry considerable interstitial albite and a little orthoclase. Superficially the rocks resemble basalts, but they differ from normal basalts in the more sodic character of their feldspar and the small amount of dark minerals; they may well be called olivine andesite. They grade into a normal pyroxene andesite.

The filling of the vesicles is varied. The most common is a soft, dull, opal-like material with shrinkage cracks and no doubt represents a gelatinous filling now partly crystallized. It is pale yellow-green [27] or a related color. In part it is amorphous and in part it is made up of fibers projecting from the walls of the cavities. These fibers have a mean index of refraction of about 1.60, a rather strong birefringence, and a positive elongation. The material is probably nontronite, celadonite, or a related mineral.

A very few minute prisms with the following optical properties were found in this material: $\beta = 1.62$; 2V large; birefringence = about 0.01, probably optically $-$. On one face it shows an extinction angle (Z$\wedge$ elongation) of 24°. On the other face it shows sensibly parallel

---

[27] Ridgway's deep seafoam-green (27''d).

extinction and the emergence of X on the edge of the field. It is therefore probably monoclinic. It is faintly pleochroic, with Z = pale greenish yellow, Y = pale orange, and X = pale pinkish orange. These properties do not determine the mineral.

Calcite is a rather abundant mineral of the cavities, as are also analcite and a number of undetermined zeolites whose optical properties are indicated below:

Cristobalite (?): Occurs in spherulites or radiating prisms with the following optical properties: $\alpha=1.480$, $\gamma=1.485$, 2V=small to medium, optically $-$, $Z\wedge$ elongation=about 43°. It is very abundant in spherulites of radiating prisms attached to the wall in the cavities of one of the andesites and is the only mineral in the cavities.

Undetermined zeolite A: Sparsely present as snow-white radiating hairlike fibers. $\beta=1.477$, birefringence barely perceptible, elongation $-$, extinction sensibly parallel.

Undetermined zeolite B: Aggregates of minute shreds. $\alpha=1.51$, $\gamma=1.52$, elongation+.

Undetermined zeolite D: Colorless to white grains. $\beta=1.505$, birefringence=0.01 about, 2V small, optically+, dispersion $\rho < v$ rather strong, extinction sensibly parallel.

The lowest flow in Windy Gulch is a biotite-hornblende andesite. It carries rather abundant phenocrysts of plagioclase and less of brown hornblende and biotite. The plagioclases have the average composition of sodic labradorite; the larger crystals carry abundant peripheral inclusions of glass. The hornblende is partly or completely resorbed by the magma, leaving the usual skeleton of magnetite grains and augite. The groundmass carries abundant laths of andesine feldspar, which grade into the larger phenocrysts, in a very fine matrix that contains rather abundant grains of magnetite and augite with an indeterminate material which is probably largely quartz and orthoclase. Tridymite is rather abundant in the small vesicles. The rock is closely related to the quartz latites.

The rock that crops out just below the mouth of Deerhorn Creek on the east approach of the bridge across West Willow Creek and in the two prospects to the west is a biotite-pyroxene andesite or quartz latite. It carries rather abundant large phenocrysts of plagioclase, augite, resorbed biotite, and magnetite. The groundmass is made up of plagioclase laths embedded in a felted to micropegmatitic intergrowth of quartz and orthoclase, specked with iron oxide. Chloritic aggregates were probably derived from small hypersthene prisms. The plagioclase phenocrysts are zoned and range from calcic to sodic andesine; the smaller laths of the groundmass are somewhat more sodic. The rocks are intermediate between the andesites and the quartz latites.

ALTERATION AND TOPOGRAPHY.

The alteration of the olivine to iddingsite and the deposition of the minerals in the amygdules are believed to have taken place while the lavas were still hot, and decomposition of the minerals of the rock by weathering has not extended to any considerable depth. The breaking up of the rock has been largely physical.   The andesite is comparatively soft and occurs in thin beds but is much more resistant than the overlying tuff.   Its upper surface is therefore commonly a bench; its slopes are characterized by a succession of benches, formed by the successive flows.

QUARTZ LATITE TUFF.

GENERAL CHARACTER AND OCCURRENCE.

At the base of the series of quartz latites that overlie the andesite there is a considerable thickness of light-colored tuff which includes local thin flows.   The greater part of the material is sandy, with scattered pebbles which are rarely over a few inches across; a part is very fine grained and a very little is conglomeratic.   The coarser material is thick bedded, but some of the finer material has very well developed closely spaced laminations.

On the ridge between Deerhorn and West Willow creeks the tuffs carry a larger amount of flow rock, the tuff-breccia is more chaotic than elsewhere, and the rocks are more commonly rhyolitic.   They probably represent a lower, local deposit in a valley.   The base, in this area, is a thinly laminated tuff not over 100 feet thick.   This is overlain by an irregular flow of rhyolite-latite, locally several hundred feet thick, and this in turn is overlain by a chaotic aggregate of thin flows and tuff-breccia.   On the west slope of the ridge the lower part of this aggregate is a glassy flow about 50 feet or more in thick-ness with a highly irregular base.   The extreme northern portion of the body is in part tuff but contains a considerable amount of massive rock in small irregular bodies.   The central portion of the body under and north of the 11,406-foot peak is in large part white tuff and latite breccia, with subordinate massive rock.   South of this peak and beginning near its top is a fan-shaped body of massive rock, with the handle of the fan forming the narrow ridge just below the peak and its broad part on the ridge to the south, where it directly overlies the rhyolite-latite.   This fan-shaped body is a steeply dipping flow which overlies successively the tuff and the rhyolite-latite.   Its outline is indicated on the geologic map (Pl. II).

The quartz latite tuff overlies a rather irregular surface of the andesite north of Bulldog Mountain.   Much of the tuff west of West Willow Creek is covered with glacial and landslide débris, and even where a continuous deposit of Quaternary material does not cover the bedrock exposures are almost entirely lacking and the mapping is necessarily unsatisfactory.

Beneath Nelson Mountain the usual tuff overlies in part the rocks of the Alboroto group, in part the Mammoth Mountain rhyolite.

North and west of the area mapped this tuff is very prominent and with the overlying flows of quartz latite forms prominent cliffs and benches for many miles north of Bristol Head.

THICKNESS.

The top of the tuff is fairly regular, but the base is locally rather irregular. Its thickness therefore varies considerably. Locally it is less than 300 feet thick, but east of Nelson Mountain it measures over 500 feet, and west of the Park Regent mine and west of Deerhorn Creek it may be even thicker.

PETROGRAPHY.

The tuff is commonly nearly white, with a drab or buff cast. The finer material is made up of fragments of glass with some broken crystals of quartz, feldspar, and biotite. The sandy part contains a large number of 1-millimeter crystals in a finer matrix of glass fragments with some larger fragments of felsitic rock. The crystals are quartz, orthoclase, plagioclase, biotite, hornblende, and the usual accessory minerals. In quantity, size, and mineral character these crystals closely resemble those of the overlying flows, to which the tuff as a whole has a close similarity. The larger fragments, which make up the main part of the tuff only on the ridge west of Deerhorn Creek, are chiefly pumice, but locally fragments of thin, platy fluidal rhyolite are abundant, and those of other rhyolites and quartz latites and of black glass are present. Andesitic rocks are rare.

MASSIVE ROCK IN THE TUFF.

Massive flows make up a part of the material called quartz latite tuff; their form and extent are shown approximately on Plate II. A prominent flow occurs on the slopes east of Nelson Mountain and extends from the eastern landslide to and beyond the east edge of the area mapped. It is about 100 feet thick and lies just below the 11,000-foot contour. The rock is Quaker drab[28] in color. It is dense and shows scattered crystals of glassy orthoclase, white andesine, and golden biotite in a felsitic groundmass. Under the microscope the groundmass is indistinctly polarizing and probably contains a considerable amount of glass. Lenses and streaks of the groundmass are somewhat coarser in crystallization and are made up of orthoclase crystals in a matrix that has a very low index of refraction and an indistinct birefringence. It may be secondary opal. A few crystals of quartz, brown hornblende, and augite are present, and apatite, zircon, and magnetite are accessory. This rock is quartz latite and approaches a rhyolite in composition.

---

[28] Ridgway's 1$''''$.

The rock of the flow west of the Amethyst mine has rather abundant kaolinized plagioclase, some glassy orthoclase, considerable pale-brown biotite, a little embayed quartz, and accessory apatite, zircon, and magnetite. The groundmass is a glass with incipient crystallization. Rather abundant gas cavities carry drusy crystals of orthoclase, quartz, and tridymite.

The greatest flow forms the slopes on both sides of West'Willow Creek above the mouth of Deerhorn Creek. The rock is light drab [29] or a nearly related color. It is commonly dense and shows rather abundant crystals of white plagioclase, as much as 3 millimeters across, and a few of glassy orthoclase and black biotite in an aphanitic groundmass. A poorly developed streaking or banding is commonly present.

The thin sections show that the phenocrysts make up considerably less than half of the rock. Plagioclase having the composition of andesine is the most abundant; embayed quartz, orthoclase, partly resorbed biotite, yellow titanite in large crystals, and the usual accessory minerals are also present. In some specimens the groundmass is very finely crystalline and is largely spherulitic, with perhaps some glass. In another section it is made up chiefly of rather large, irregular rounded blotches that can be seen only in polarized light and are very fine intergrowths of quartz and orthoclase. They are embedded in an exceedingly finely crystalline matrix and are packed closely together in some parts of the section but scattered, irregularly in other parts. As seen in ordinary light this ground has a rather prominent fluidal streaking, and the streaks pass through the polarizing blotches and the finer matrix indiscriminately. The ground is everywhere filled with red trichites of iron oxide. In composition the rock is near the border between the rhyolites and the quartz latites and may properly be called a rhyolite-latite. This rock offers little resistance to weathering and disintegrates into small hackly fragments. Outcrops are poor, and most of them are made up of the disintegrated rock. The slopes are commonly smooth and rather steep and show the disintegrated rock in place very near the surface.

The massive rock associated with the tuff that overlies this rhyolite-latite west of Deerhorn Creek is in large part a black or dark-gray glass; rarely it is nearly white. It carries rather abundant inconspicuous crystals of glassy, striated felsdpar as much as 5 millimeters across and some biotite. Microscopically the rocks show also a few crystals of orthoclase, quartz, and titanite, with accessory zircon, apatite, and magnetite. Many of the rocks have also green hornblende and augite. The titanite is abundant in 1-millimeter

---

[29] Ridgway's light purple-drab (1''''b).

crystals. The plagioclase is zoned andesine. The groundmass of the glassy rocks shows incipient crystallization, chiefly along the cracks. In some specimens abundant red rounded botyroidal spherulites several millimeters across are embedded in a greenish glass. These flows have not been separated from the associated tuff on the geologic map.

The rock of the mapped fan-shaped flow south of the 11,406-foot peak west of Nelson Mountain is dark purple-drab [30] and superficially resembles some of the dense, fine-textured quartz latites of the Piedra group. It carries abundant 1-millimeter crystals of andesine, with some of quartz, orthoclase, biotite, augite, and titanite and accessory apatite and magnetite. The groundmass is fluidal, in part spherulitic, and in part very finely crystalline. The rock is a quartz latite.

WEATHERING AND OUTCROPS.

This tuff is but little consolidated and is relatively soft, being much less resistant than the underlying or overlying rocks. The overlying quartz latites in particular are thick, cliff-forming flows, and the soft tuff gives way under their load, thus giving rise to the landslides which so commonly cover the tuff. In general the tuff forms comparatively gentle slopes with few outcrops, and there is commonly a more or less well-developed bench near its base. In Plate IV, *A* (p. 3), typical slopes of the tuff are shown at the rear of the mines from the center of the photograph to the right. On the extreme right the gentle slope below the cliffs is formed of the tuff, modified by glaciation and landslides. Locally, as on the southeast slopes of Nelson Mountain, the tuff forms steep slopes with good outcrops. These white outcrops with their striking castellated forms are locally called the "white elephants." Where flows are present in the tuff they form a succession of benches and cliffs. The white color of the exposures of the body west of the mouth of Deerhorn Creek and the extreme irregularity of the complex has led the local prospectors to call the mass "The Blowout."

RAT CREEK QUARTZ LATITE.

GENERAL CHARACTER AND DISTRIBUTION.

Immediately overlying a fairly regular surface of the quartz latite tuff and closely related to it is a series of lava flows with some interbedded tuff, here called the Rat Creek quartz latite, from their development on Rat Creek. The flows are mostly thin and rather irregular; the tuff forms a minor part of the material. The several flows and the interbedded tuff consist of quartz latites of very similar character; they differ but little from the material that makes up the

---

[30] Ridgway's 1''''i.

underlying tuff and from the overlying Nelson Mountain quartz latite.

This latite underlies the Nelson Mountain quartz latite on the ridge between Rat and West Willow creeks, but owing to the large amount of landslide material beneath the cliffs of the Nelson Mountain quartz latite, it is shown as a number of disconnected outcrops. It is also present on the slopes of Nelson Mountain.

THICKNESS.

Although there is no marked irregularity either at the base or top of the Rat Creek quartz latite, it varies considerably in thickness. Between Rat and West Willow creeks it is only 150 feet thick at the southern exposures but becomes considerably thicker to the north. On the slopes of Nelson Mountain it is from 400 to 500 feet thick.

PETROGRAPHY.

Megascopic features.—In color the rocks are pale drab.[31] They are rather dense and have fairly well developed flow structure shown by horizontal planes of fracture but not by prominent banding. Phenocrysts 1 or 2 millimeters in cross section make up about half the rocks. They are chiefly white to moderately clear striated plagioclase; there is a less amount of clear, glassy quartz and orthoclase, considerable biotite in the usual black flakes and hornblende in black cleavable prisms. Light-green augite is abundant in most of the rocks, and yellow-brown titanite can be seen in some. The groundmass is aphanitic. Inclusions of quartz latites and rhyolites much like the body of the rock are abundant in some of the flows.

Microscopic features.—The microscopic examination shows that the phenocrysts are somewhat diverse in size and are commonly broken. Plagioclase is the chief; embayed quartz and orthoclase are never abundant and are not always present; biotite is always present; both hornblende and augite are commonly present, but some flows contain only one of them; sphene is abundant and occurs in large crystals in some of the rocks; magnetite, apatite, and zircon are accessory; and tridymite was found in a few of the sections. The plagioclase shows the usual zonal growths and ranges from oligoclase to oligoclase-andesine. The hornblende in most of the flows is a chestnut-brown variety but in some is a green or olive-green variety. Both it and the biotite have been greatly resorbed by the magma. The groundmass consists chiefly of spherulitic growths of orthoclase; in some specimens it is partly glassy. Tridymite and cristobalite are present in rounded aggregates or intergrown with the orthoclase of the spherulites. Streaks and irregular areas of coarser crystallization are common in most of the specimens.

---

[31] Ridgway's pale purple-drab (1''''*d*), pale Quaker drab (1''''''*d*), or a nearly related color.

WEATHERING, OUTCROPS, AND TOPOGRAPHY.

Most of the Rat Creek rock is more resistant than the underlying tuff but less so than the overlying flow. Its outcrops are rather poor, and it forms rather steep slopes between the gentle slopes of the tuff and the cliff of the overlying flow. Much of it is covered by landslide.

NELSON MOUNTAIN QUARTZ LATITE.

GENERAL CHARACTER AND DISTRIBUTION.

The upper flow of the Piedra group in the Creede area is a regular flow of quartz latite of uniform character. It makes the nearly continuous upper cliff and caps the divide east of Rat Creek, in the northeast corner of the area. It is also the cap rock on Nelson Mountain, whence it is named the Nelson Mountain quartz latite. A similar flow overlies the tuff and forms the mesa north of Bristol Head. It commonly forms prominent mesas, nearly surrounded by cliffs.

THICKNESS.

The top of this flow no doubt corresponds approximately to the comparatively flat tops of the mesas. The flow is fairly uniform in thickness and between Rat and West Willow creeks it is probably 350 feet thick; on Nelson Mountain the thickness is less than 200 feet.

PETROGRAPHY.

Megascopic features.—In color the rock is rather uniformly red-brown.[32] The phenocrysts are about 1 millimeter in cross section and are about equal to the groundmass in amount. They consist chiefly of porcelain-white striated plagioclase, with rather abundant black plates of biotite and prisms of black hornblende and of pale-green augite; glassy orthoclase and quartz are less conspicuous. The groundmass is aphanitic and rather dense; fluidal structure is present but not conspicuous.

Microscopic features.—A detailed microscopic study showed that the plagioclase has the average composition of andesine. The hornblende is of a brown variety, and both it and the biotite have been somewhat resorbed by the magma. The quartz phenocrysts occur in small amount and are embayed. The accessories are titanite, in rather large and abundant crystals, magnetite, apatite, zircon, and hematite. The hematite occurs in very minute specks and shreds as a pigment to the groundmass. The groundmass is very finely crystalline and is largely spherulitic; it is made up chiefly of orthoclase and quartz. Streaks and bunches of coarser crystallization occur; a little tridymite is present in some of these coarser streaks. The rock is not very different from some of the quartz latites of the Alboroto group.

---

[32] Ridgway's purple-drab (1'''') or a slightly lighter tint or darker shade.

WEATHERING AND OUTCROPS.

The minerals of this rock show little alteration, and the weathering is largely mechanical. The rock breaks into flat plates along the fluidal planes and on the tops of the ridges yields a fair amount of soil. As it is a very resistant and rather thick flow and overlies softer tuffs and thin lava flows, it breaks down largely by landslides of various sizes, thus causing a recession of its cliffs. The flow forms mesas at its upper surface, surrounded by almost continuous cliffs, below which are commonly talus heaps and landslide débris. The mesas of this rock capping Nelson Mountain are shown against the sky line at the left of the view given in Plate IV, B (p. 3); those in the northwest corner of the Creede area are shown against the sky line at the extreme right of Plate IV, A.

Chapter V.—CREEDE FORMATION.

NAME, GENERAL CHARACTER, AND OCCURRENCE.

A considerable thickness of rhyolitic tuffs and bedded breccia, which contain local bodies of travertinous spring deposits in their lower part and some intercalated lava flows in their upper part, accumulated in a deep valley or basin that had much the same character and position as the present valley of the Rio Grande from Wagonwheel Gap westward to Trout Creek, a distance of about 25 miles, but was considerably narrower. The name Creede formation is here proposed for this deposit, from its extensive and characteristic development on the slopes on both sides of Willow Creek about the town of Creede.

In general the Creede deposits occupy an area in which the Hayden map [33] represented Green River Eocene beds with Carboniferous limestone above them on the slopes. The Hayden reports [34] contain practically no reference to these deposits in the Rio Grande valley, and it seems plain that the Creede tuffs in the bottom of the valley were called Green River Eocene, while the interbedded travertine deposits that exhibit massive outcrops in some places were referred to the Carboniferous without any good ground. Collections of plant remains from the lower tuff beds of the Creede formation indicate that these beds should be correlated with the Florissant lake beds (Miocene).

The character of the basin in which the Creede formation was deposited is well shown by the way in which the beds commonly lap up on the steep slopes of the older basin of deposition made up of similar rocks, which form the abrupt slopes of to-day. This is also shown by the geologic map and the structure sections (Pl. II). The Creede beds occupy the floor of the valley of the Rio Grande and are surrounded by steep slopes of the older rocks, which commonly rise 3,000 feet or more above the valley in a distance of a few miles. This relation is not due to folding or faulting but to the fact that the Creede formation was deposited in a basin that was deeper than that of the present Rio Grande valley and probably had steeper walls. The resulting irregularity at the base of the Creede formation is well shown on both sides of Creede; to the east the base of the formation rises nearly 2,000 feet in elevation within less than a mile. (See Pl. II.) In Plate IX the base of the Creede formation is shown at the top of the cliffs on the right over the buildings.

---

[33] U. S. Geol. and Geog. Survey Terr. Atlas, sheets XV, XVII, 1881.
[34] U. S. Geol. and Geog. Survey Terr. Ninth Ann. Rept., pp. 153–157, 1877.

STRUCTURE AND THICKNESS.

In general the tuff beds dip gently away from the mountains toward the south and southeast, and the dips are commonly steeper near the borders of the body. Dips of 10° or even 20° are common near the borders, and much steeper dips occur in places. Away from the borders the dips are low, but accurate estimates of the average dip for any considerable area are difficult on account of the poor exposures and local undulations. The general structure in the tuffs of the valley is that of a gentle syncline. It is not known how much of the dip is due to tilting since deposition, as beds laid down rather rapidly in a comparatively small basin surrounded by very steep slopes might have a considerable inclination at the time of deposition, and this inclination would probably be steeper near the borders of the basin.

No satisfactory estimate of the maximum thickness of the Creede beds can be made, as the structure is uncertain, the top is nowhere preserved, and the base is exposed only on the steep sides of the basin. About 500 feet of the lower member is exposed on both sides of Willow Creek below Creede, and neither the top nor the base is shown; just east of Creede, where the top of the lower member is preserved, about 800 feet is exposed. Approximately 1,000 feet of the upper member is exposed east of Windy Gulch. In the middle of the valley the thickness was probably considerably greater.

SUBDIVISIONS.

The Creede formation has been subdivided on the map into three lithologic units, which, however, have no sharp lines of separation. The lower member is made up entirely of fragmental material, the greater part of which was deposited by water, although some of it represents talus and similar accumulations. Much of this material is a thinly laminated white shaly tuff; part is sandy, and part is breccia and conglomerate. The material is entirely of igneous origin, and a greater part of the larger fragments were derived from the rocks of the Alboroto group, whereas the finer tuffs were deposits of volcanic ash from local eruptions.

Interbedded with the tuff of the lower member are beds of travertine, which was deposited by springs, probably hot springs, at the time the tuff was being laid down and at various horizons. It is in part a surface deposit, in part a deposit in the channels through which the waters came, and in part a deposit formed by cementation and replacement of the tuffs. It occurs in very irregular bodies whose contacts are not everywhere sharply marked. Only the larger bodies are shown on the map, and their boundaries as indicated are more or less diagrammatic.

The material of the upper member is somewhat coarser than that of the lower member; it is made up mainly of rather well-bedded breccia, conglomerate, and tuff. A considerable part of the fragments consists of a kind of rhyolite that is found only here and in the associated thin intercalated flows.

LOWER MEMBER.

General character.—The lower member of the Creede formation is made up in large part of fine-grained and well-bedded material, which is commonly creamy or light grayish. In part it is very thin bedded and shaly and closely resembles a siliceous shale, although composed almost wholly of volcanic material. Much of this tuff carries plant remains, some of which are well preserved. Thicker beds of sandy material and beds or lenses of conglomerate are present. Most of the pebbles are well-rounded fragments of igneous rock; some are subangular or even angular. Near the contact with the underlying rocks coarse and angular material is much more abundant, and at the base of the original steep slopes of the older rocks the local material of the Creede beds really represents an indurated talus from the cliffs. In the upper part of the member, west of Rat Creek, the material is more thickly bedded and much of it is sandy or conglomeratic.

Petrography.—The shaly tuff is rather uniform in character and is made up in large part of fragments of rhyolitic glass with an occasional fragment of feldspar, quartz, and biotite. It is no doubt an ash from a near-by volcanic vent. The sandy beds carry very abundant crystals of orthoclase, plagioclase, quartz, and biotite, rare fragments of other minerals, and a varying number of fragments of felsitic or pumiceous rhyolite.

The pebbles of the conglomeratic part of the tuff consist chiefly of fragments of the underlying rocks but include also a great variety of other rocks, chiefly rhyolites and quartz latites. The fragments of the conglomerates and breccias about Creede were derived chiefly from the Willow Creek and Campbell Mountain rhyolites but in part from rocks similar to the quartz latites of the Alboroto group. Large angular blocks of tridymite latite similar to that of the Piedra group are present just east of Creede. The nearest exposures of the tridymite latite are much higher on the slopes on Bulldog Mountain or in the high mountains a few miles to the northeast. The fragments may well represent landslide or talus blocks from cliffs of this rock that have since been removed by erosion. A few fragments of a variety of other rhyolites, quartz latites, and andesites that can not be correlated with any of the rocks of the vicinity are also present, and these are more abundant in the thin lenses of conglomerate that occur in the thin-bedded tuffs at some distance from the mountains. In other places, as near Wagonwheel Gap, the fragments in the tuff are chiefly quartz latites similar to those that formed the adjacent slopes at the time of deposition.   Where the fragments are angular, especially near the contacts, the greater part are identifiable as belonging to the immediately adjacent rocks.   In the Iowa tunnel and elsewhere near the contact with the lower rocks the material is made up of somewhat indurated small angular fragments of the Willow Creek rhyolite and no doubt represents simply an accumulation of talus at the base of the steep slopes of the rhyolite during the Creede epoch.

In general both the tuff and conglomerate are fairly well indurated and are much harder than the somewhat similar tuff of the Piedra group.   Near the travertine bodies the tuff is filled with calcite and much of the coarser material is cemented by reddish-brown or yellowish hydrated iron oxide or by a green ferruginous material. Locally, especially in the upper part, the tuff and breccia are cemented by silica.

INTERBEDDED TRAVERTINE.

Character and distribution.—Within the tuff of the lower member of the Creede formation are a large number of bodies of travertine. The largest of these is just east of Sunnyside Creek, where it forms prominent cliffs; other bodies are mapped between Creede and Sunnyside and between Creede and Dry Gulch.   These travertine bodies are also characteristic accompaniments of the Creede formation beyond the Creede area.   Large bodies are present in the lower part of the Lime Creek basin near the western extremity of the Creede formation, between Fir and Shallow creeks, in the drainage basin of Farmers Creek, near Wagonwheel Gap, and in the drainage basin of Deep Creek.   They are rather more abundant near the borders of the Creede formation than in the center   The bodies are very irregular in form, and their contacts are in places indefinite; consequently the mapping is much generalized and is intended to show only the approximate form and location of the larger bodies.   A great part of the travertine is a light-grayish deposit of very fine grained calcite, in some places very dense, in others rather porous or highly cellular.   There is commonly a considerable amount of limonite, especially in the more porous variety, and some of the material is highly ferruginous. Locally, as in parts of the body north of the 9,208-foot hill wes of Willow Creek, there is a light, porous deposit of gypsum, stained and coated with limonitic material.   Part of the travertine incloses fragments of rock.   Locally it carries much chalcedony, with some quartz, filling original cavities or as veinlets, and in places it contains deposits of siliceous sinter.

Origin.—The travertine and less common siliceous sinter were deposited about the openings of springs, probably hot springs, which must have been very abundant about the borders of the old basin.

The travertine is usually mixed or interbedded with the fine-grained tuff, and in some places it appears to act as a c ment for the tuff; in part it forms lenses or less regular bodies in the tuff. Some of the travertine was probably deposited in the conduits of the springs; the strip through the middle of the body east of Sunnyside, which forms the pronounced outcrop of this body, is believed to have been so deposited. It forms a dikelike outcrop striking a little north of east above the 9,100-foot contour, with very steep walls, over 100 feet high in places, on the south side and a bench or low depression on the north side. It is almost entirely travertine, but the rest of this body, included in the travertine on the map, carries much tuff intimately mixed with the travertine.

Travertine deposits, somewhat similar to those in the Creede formation, have been formed elsewhere on a large scale about the borders of interior undrained lakes, by the precipitation of calcium carbonate from the waters of the lakes. The abundance of silica and iron in the travertine of the Creede formation and the presence of bodies of cellular gypsum and limonite, together with the small size and irregular form and distribution of the bodies, make it seem unlikely that this travertine was so deposited.

UPPER MEMBER.

General character and distribution.—The upper member of the Creede formation is made up of considerably coarser material than the lower member, and the material is better sorted, more distinctly bedded, and more rounded than the talus-like material that constitutes most of the coarse part of the lower member. In addition to fragments of the older rocks from the near-by slopes it carries abundant fragments of rhyolites found only in this member, and some of the conglomerates are made up entirely of such rhyolites; thin flows of similar rhyolites are interbedded with these conglomerates.

The two members are not sharply distinct, and the contact has been drawn more or less arbitrarily. It was intended to include in the upper member all the pebble beds that are made up of rocks similar to those of the intercalated flows.

This member is best exposed in the drainage basin of Windy Gulch and on the slopes west and south of the Commodore mine. The contacts of this body are fairly well mapped except where it is in contact with the rocks of the Piedra group, for there the exposures are very poor and the contacts and relations are uncertain. The flats near and northwest of Bachelor are included in this member, although much of the material on them may be Quaternary wash. The map also shows an area on the south slope of Mammoth Mountain, but this area is almost entirely lacking in exposures and the contacts are very greatly generalized. The typical rock was seen only in a few prospects at an elevation of about 10,550 feet just west of the isolated cliffs of Willow Creek rhyolite.

Petrography.—The clastic rock varies greatly in the degree of sorting and rounding of the fragments, the coarseness of the material, the petrographic character of the fragments, and the amount of induration. On the whole it is considerably coarser than the lower member of the Creede formation, is somewhat more indurated, and carries abundant fragments of rock similar to the associated flows. A small part of the material is rather thin bedded and of fine shaly texture, somewhat more is sandy, and the greater part is conglomerate or breccia.

A large part of the material is rather well bedded and sorted and consists of fragments as much as several inches across, generally not well rounded, in a large amount of matrix. The fragments consist chiefly of the Mammoth Mountain rhyolite, with some of the Willow Creek rhyolite, but locally, west of Windy Gulch, they consist almost entirely of rocks similar to the associated rhyolite flows, especially of the banded type of massive rock. (See p. 67.) In a few places rocks resembling the tridymite latite and more or less similar to the rhyolite breccia of the Piedra group are present; other rocks are rare. The matrix is dark red or drab, rarely white, and is very fine, hard, and dense, except for occasional cavities, and the rock superficially resembles a flow breccia. The cementing material is chiefly quartz and chalcedony, with considerable iron. Barite and jarosite were found in a number of places, chiefly in the cavities; they do not appear to be confined to the vicinity of the veins, as they are common in Windy Gulch. A large part of the breccia has apparently been cemented by ferruginous quartz, with some sulphates.

The fine-textured layers are in large part indurated to hard, flinty rocks, in most places stained red or drab by iron. They carry a few poorly preserved fragments of fossil plants. The sandy layers in the main consist either of broken crystals of feldspar and biotite or of fragments of felsitic rhyolite and have a cement of ferruginous quartz.

The coarser beds of conglomerate, which are interbedded with the lava flows in Windy Gulch and near the Commodore mine and are also present in lenses in the overlying breccia, are made up of well-rounded pebbles, the largest a foot or more across, with a moderate amount of sandy matrix. They are usually cemented by silica and iron compounds to a hard rock. Most of the fragments are rock like that of the associated flows; some are fragments of a rock with more prominent phenocrysts, and the pebbles of this type are much altered.

Massive rocks.—In the lower part of the section in the drainage basin of Windy Gulch and to the northeast as far as the Bachelor mine a number of thin flows are present and alternate with beds of conglomerate made up of fragments of rock similar to that of the flows. They have not been separated on the geologic map, as they are local, thin, and poorly exposed and are closely associated with the conglomerates, which in turn grade into the normal breccia.

The greater part of the flows is made up of a rather porous drab [35] rock which carries phenocrysts from 1 to 2 millimeters in cross section, of glassy orthoclase, white porous feldspar, and dark-brown plates of biotite. These phenocrysts make up about half the rock and are embedded in a porous, aphanitic material. The microscope reveals a few quartz grains and shows that the white feldspar is a peculiar microperthite in porous, skeleton crystals and is partly altered to sericite. This gives an imperfect wavy extinction, as the intergrowths are almost submicroscopic. Orthoclase predominates. The biotite is partly resorbed. Accessory apatite, magnetite, and zircon are sparsely present. The groundmass shows beautiful wavy, fluidal bands of lighter tone and coarser crystallization. In part the crystallization is submicroscopic, but irregular streaks and bands are made up of closely packed spherulites, here minute, there large. The rocks are fresh except for the presence of a little secondary calcite and sericite. The albite may have been derived from a more calcic feldspar. In one of the sections the phenocrysts have very ragged borders due to their influence in orienting the adjoining groundmass. The groundmass is granophyric and rather coarsely though very irregularly crystalline. Just south of the Commodore mine one of the flows shows white blotches, which can be recognized with a pocket lens as areas of radiating fibrous spherulites.

Another type of massive rock found only west and south of the Commodore mine carries rather numerous rough cavities 1 inch in largest dimension. It has fewer phenocrysts than the type described above and a much finer groundmass but is otherwise similar.

South of Bulldog Mountain a third type is exposed partly as a massive rock and makes up much of the breccias. It is a pale purple-drab rock with rather broad bands and lenses of lighter tone. It carries rather abundant crystals of glassy orthoclase and a few of white feldspar and dark-brown biotite. The microscope shows that the plagioclase is near albite and is somewhat altered to sericite. The orthoclase crystals include well-formed crystals of albite. The groundmass of the broad dark-colored bands is very finely crystalline; that of the light bands is rather coarsely crystalline, with the quartz and orthoclase intergrown. The fibers or prisms of feldspar tend to grow out from the walls, and the quartz fills in between them. All the flows are rhyolites rather rich in soda.

---

[35] Ridgway's Quaker drab (1′′′′′) to purple-drab (1′′′′).

WEATHERING AND TOPOGRAPHY.

The beds of the Creede formation are by far the least resistant of all the large bodies of rock within or near the Creede area, and they have had a marked influence on the development of the topography. The broad valley of the Rio Grande, with its comparatively gentle contiguous slopes from Trout Creek to Wagonwheel Gap, a distance of about 25 miles, corresponds to the area occupied by these tuffs and is in marked contrast with the steep, rugged slopes surrounding the valley and with the canyons of the Rio Grande both above and below. Equally marked is the contrast between the rugged rock canyon of Willow Creek above Creede, where it is cut in the Willow Creek rhyolite, and the broad valley and rolling hills below Creede, where it is eroded in the Creede formation. Plates III and XI show the low, rolling grass-covered hills of tuff about Creede and the marked change in topography above Creede where the Willow Creek rhyolite comes out from under the tuffs.

ORIGIN.

The greater part of the clastic material of the Creede formation was clearly deposited by water. The conglomerate was probably laid down in part along the shore of the lake, in part by contributory streams during freshets. Near Creede conglomerate is much more abundant in the upper part of the formation and makes most of the clastic material in the upper member; but this may not be true in other portions of the basin. Indeed, the upper member near Creede is probably largely fluviatile material. Ancient talus and similar subaerial accumulations make up a small part of the formation. The fine shaly tuff must have been deposited in comparatively still water, and its considerable though local distribution, together with its great thickness and the form of the basin of deposition, are rather satisfactory evidence that it accumulated in a local intervolcanic lake. The form of the old basin beyond the present Rio Grande Valley is not known, as younger rocks hide it both above Trout Creek and beyond Wagonwheel Gap.

The material of this formation was derived from four sources. Much of the fine tuff represents volcanic ash which was thrown out from some unknown but near-by volcanic vent and fell in a lake or on sides of the basin surrounding it and was deposited, after sorting, in the quiet water of the lake. The greater part of the coarse material was derived by the ordinary processes of erosion from the mountains surrounding the basin and was brought into the lake by the torrential streams that fed it. This portion is made up almost entirely of the underlying rocks. The travertine, as has already been stated, was deposited by hot springs contemporaneously with the deposition of the fine tuff. Finally, a number of thin lava flows are present in the upper part of the formation.

POSITION IN THE SECTION.

The Creede formation clearly overlies the rocks of the Alboroto group. The contacts and relations are well exposed on both sides of Creede and in other places and a large part of the fragments from the coarse material in the Creede formation were derived from the two rhyolites of the Alboroto group. It also overlies the rocks of the Piedra group, but although it is in contact with Piedra rocks near Sunnyside, in the drainage basin of Windy Gulch, and near Bachelor, the exposures are nowhere sufficient to establish the relation between the two. The Creede and Piedra rocks are also in contact east of the area included in this report, but here too the relations are obscure. Although the Creede formation everywhere occupies the slope below the Piedra group, the form of the contacts indicates that the Creede was deposited in a basin with steep slopes made up of the rocks of the Potosi volcanic series; this is shown more clearly east of the area mapped, where the soft tuffs and breccias of the Creede formation abut against steep slopes of flows of the Potosi volcanic series.

On the whole the Creede breccias contain fewer fragments of rocks that can be identified as belonging to the Piedra group than would be expected from the relations, but fragments of the tridymite latite were found in the lower member of the Creede formation just east of Creede and in the upper member in Windy Gulch. Moreover, two of the formations of the Piedra group that should have contributed largely to the Creede beds, the Windy Gulch rhyolite breccia and the Mammoth Mountain rhyolite, weather into small fragments, and these could hardly be distinguished from fragments of the older Campbell Mountain rhyolite. To the east, in the drainage basin of Farmers Creek, where the Creede beds lap against large masses of rocks belonging to the Piedra group, fragments of those rocks are more abundant in the Creede formation.

In the area included in this report no rocks have been found overlying the Creede formation. However, at Wagonwheel Gap, only 8 miles to the southeast, a thick flow of a hornblende-quartz latite, which is without doubt related to the Fisher quartz latite, overlies the tuff. The position of the Creede formation is therefore between the Fisher quartz latite and the Potosi volcanic series.

AGE.

The thin-bedded tuffs of the Creede formation have furnished the only determinable fossils found in this vicinity. These tuffs commonly carry a moderate number of plant remains, some of which are well preserved. Collections were made from three localities in the Creede area and submitted to F. H. Knowlton for identification. Below are the lists of forms reported by him from each locality:

2 (5951). Creede quadrangle, Colorado. Ridge north of Pierce Creek, near Hot Spring Hotel, Wagonwheel Gap, at elevation of 9,000 feet:

Minute fragments of bark, coniferous leaves, etc., but nothing determinable.

24. (5952). South bank of Rio Grande 150 yards above wagon bridge over Rio Grande, 3½ miles below Creede, Colo.:

Fontinalis pristina? Lesquereux.
Cercocarpus antiquus? Lesquereux. A very narrow small leaf.

93 (5953). Tuff, west bank of Rio Grande one-fourth mile north of Sevenmile Bridge, 7 miles above Creede, Colo.:

Pinus wheeleri Cockerell. Leaves in fives.
Sequoia sp.?
Thuya sp.
Dicotyledon, narrow, wedge-shaped at base; no nervation.
Quercus pyrifolia? Lesquereux.
Fontinalis pristina Lesquereux.[36]

536 (6198). Same locality as 93:

Adiantites? sp. new?
Planera myricaefolia (Lesquereux) Cockerell.
Pinus wheeleri Cockerell.
Pinus sp. cone, like P. florissanti but smaller.
Carduus florissantensis Cockerell.
Vitis florissantella Cockerell.
Conifer, probably Picea or Abies, undoubtedly new.
Mahonia? sp. trifoliolate and new.
Abies, two species, new.
Juglans, new?
Celtis mccoshii Lesquereux.
Lomatia hakeaefolia Lesquereux.
Ribes protomelaenum Cockerell.
Insect, beetle?
Feather, like those of Florissant.

Mr. Knowlton makes the following comments on this flora:

These species are all or nearly all found in and are highly characteristic of the Florissant flora. The question then arises as to the age of the Florissant deposits. For many years they were supposed to be in the approximate position of the Green River formation, but on the determination of the fossil insects the beds have of late years been referred to the Oligocene. Still more recently extensive explorations have been made in the Florissant plant beds, with the result of bringing to light a great number of new and on the whole of very modern appearing forms, and the conviction has been growing that these beds had been placed too low in the scale. Extensive Tertiary floras from this country, as well as from other parts of the world, have lately become available for comparison, and it now seems probable that the Florissant beds are upper Miocene. In any event I feel perfectly justified in stating that the plants here submitted from the vicinity of Creede, Colo., are of the same age as the plant beds of Florissant, Colo., and that this age is certainly Miocene, and in all probability upper Miocene. As to the original reference of these beds to the Green River, I can only say that none of the plants submitted belong to the Green River flora, while all but one or two that have been identified do occur in the Florissant beds.

---

[36] This has been shown to be a feather and not a moss.

Chapter VI.—FISHER QUARTZ LATITE.

GENERAL RELATIONS.

In the Creede area a volcanic series still later than the Potosi is represented about MacKenzie Mountain by the great lava flow belonging to the member named the Fisher quartz latite in 1917.[37] Beyond the area mapped in nearly all directions this latest volcanic series is extensively developed. It attains a great thickness on the divides, both to the north and south of the Rio Grande, and tongues and patches of it extend toward the Rio Grande valley and locally cross the valley. It is made up of a succession of lava flows and pyroclastic deposits. Some of the flows are several hundred feet thick and persist for long distances; others are irregular and local. The pyroclastic part is largely very chaotic and contains rather abundant irregular bodies of flow rock. The rocks of this series are quartz latites, rhyolites, and andesites. They are nearly all characterized by larger and more conspicuous phenocrysts than the other rocks of the area and thus can usually be recognized with little difficulty.

The relation of this volcanic series to the Potosi volcanic series is well shown in the vicinity of MacKenzie Mountain, where the Fisher quartz latite overlies the several members of the Potosi volcanic series in an irregular manner and clearly flowed over an erosion surface of considerable relief and filled in the valleys and other irregularities. Beyond the limits of the area mapped, in the upper East Willow Creek basin and elsewhere, the irregularity at the base of this series is still more marked and is persistent wherever this series has been found.

The relation of this volcanic series to the Creede formation is not shown in the area covered by this report, as the two are nowhere in contact. However, about 8 miles to the southeast, at Wagonwheel Gap, a great flow belonging to this volcanic series rests successively on the Creede formation and rocks of the Potosi volcanic series and has a lower contact that cuts across the contours very irregularly. The hard rock that forms the gap belongs to this formation. Clearly a considerable interval of erosion succeeded the deposition of the Creede lake beds and preceded the extravasation of the rocks of this upper volcanic series.

---

[37] Colorado Geol. Survey Bull. 13, pp. 20, 23–33, 1917. Named for development on Fisher Mountain, Creede quadrangle.

GENERAL CHARACTER AND DISTRIBUTION.

On MacKenzie Mountain and the ridges to the north and south is a prominent quartz latite containing abundant crystals of green augite and large and prominent phenocrysts of feldspar and biotite.   Within the area mapped it probably represents a single flow, although a part may represent material that solidified within the vent through which the material of the flow was extruded.   As shown on Plate II, this rock occupies the large area including MacKenzie Mountain and two smaller areas on the same ridge south of the Kreutzer fault.   In addition, there is a small outcrop of this rock surrounded by glacial drift in the upper Rat Creek basin, another just east of Rat Creek, and a third west of Rat Creek and northeast of MacKenzie Mountain. The last-mentioned body, which caps the 10,800-foot hill, is poorly exposed and may represent slide rock from the cliffs above, although it is believed to be in place.   Similar rocks, which no doubt represent the same eruptive series, although not the same flow, are extensively developed on the higher slopes both north and south of the Rio Grande.

THICKNESS.

Wherever observed this flow came over a surface whose relief was comparable with that of the present surface, and within the Creede area the extreme irregularity of its base is well shown.   On the east side of the saddle just north of MacKenzie Mountain the base of this flow is well exposed at an elevation of 11,150 feet; thence to the southeast the base of the cliff, which is practically at the base of the flow, as is indicated by the glassy rock at this horizon, cuts sharply across the contours and within an eighth of a mile crosses the 10,850-foot contour, a fall of 300 feet.   Just north of this locality the steep slope of the base is even more striking, as it is also on all sides of the isolated body on the east side of Rat Creek.   The latter body evidently rests on very steep slopes of the older rocks, and it was at first believed to be intrusive, but the identity of the rock with that of the flow across the creek and the glassy layer at its base, together with the details of the contacts, where exposed, show that the body is a flow resting on the rocks of the Piedra group.

Owing to the extreme irregularity at the base of this flow its thickness varies greatly.   Moreover, the top is nowhere certainly exposed, although the small plateau northwest of MacKenzie Mountain is believed to represent nearly the top of this flow.   It is commonly several hundred feet thick and locally as much as 500 feet.

PETROGRAPHY.

Megascopic features.—The rock is fairly uniform in appearance and varies chiefly in color, porosity of the groundmass, and the amount of glass in the groundmass. The color [38] is purple-drab in the denser rock to light Quaker drab or light mouse-gray in the more porous rocks. The glassy layer at the base is deep olive-gray.[39] Except for the basal glassy layer, all the rocks show a few small gas pores, and some of the rocks are decidedly porous. The flow is characterized by the number and especially the size of its phenocrysts, which about equal the groundmass in amount and are commonly from 2 to 4 millimeters across; a few measure 1 centimeter or more. They are chiefly porcelain-white plagioclase, with rather prominent black flakes of biotite and prisms of pale-green augite. The groundmass is aphanitic.

Microscopic features.—In addition to the minerals recognized with a pocket lens the microscope shows a few crystals of zircon, apatite, and magnetite. The groundmass is largely submicroscopic in crystallization and is rhyolitic in character. The plagioclase is andesine-labradorite in composition and carries central inclusions of the other constituents. The biotite is considerably resorbed, and the augite is somewhat altered to chloritic material. A little tridymite is present in the porous parts of some of the specimens.

WEATHERING AND OUTCROPS.

The rock is little altered; it is very hard and resistant to weathering and nearly everywhere forms prominent cliffs. It is underlain by softer rocks, and landslides have been important factors in breaking up the flow. The slopes beneath its cliffs are commonly covered with a heavy mantle of landslide material from this cliff-forming quartz latite.

---

[38] Ridgway's 1'''' to 1'''''b or 15'''''b.
[39] Ridgway's 23'''''.

Chapter VII.—INTRUSIVE ROCKS.

TYPES.

In the Creede area intrusive rocks are neither great in extent, numerous in individual bodies, nor varied in character. Four types of these rocks have been recognized—rhyolite porphyry, quartz latite porphyry, basalt, and hornblende-quartz latite porphyry. The rhyolite porphyry is fairly uniform in character and was probably all intruded at about one time, during the later part of the period represented by the Alboroto group. It occurs south of Bulldog Mountain and east of Willow Creek in irregular or sill-like intrusions, none of which are extensive. The quartz latite porphyry is very uniform in character and no doubt was all formed at about the same time. It is confined to the southeast corner of the area and occurs in dikes as much as a few hundred feet across, striking west of north. It is probably of about the same age as the Fisher quartz latite and may be the intrusive equivalent of that or a related flow. The basalt was found in only one narrow dike in the lower Rat Creek basin and cuts the lower part of the Piedra group. The hornblende-quartz latite porphyry was found only in the mine workings.

RHYOLITE PORPHYRY.

GENERAL CHARACTER AND DISTRIBUTION.

A nearly white rhyolite porphyry characterized by large, glassy crystals of orthoclase is present in several bodies of considerable size south of Bulldog Mountain and in some smaller bodies on the slopes west of Willow Creek just above Creede. Small bodies of this rock were found in some of the underground workings, and a somewhat similar rock is exposed on both sides of West Willow Creek just below Weaver. The details of the distribution are shown on Plate II.

This rhyolite porphyry is very soft and gives few good outcrops. South of Bulldog Mountain, in particular, outcrops are almost entirely lacking, and the boundaries were mapped largely from the appearance of the soil and talus. The character of the contacts was rarely observed, and the mapping is therefore generalized.

RELATION TO ADJOINING ROCKS.

Wherever exposures are sufficient to offer any data as to the relation of this rhyolite porphyry to the adjoining rocks an intrusive origin is strongly suggested. It nowhere appears to occupy a definite horizon in the section, although it is confined to the rocks of the Alboroto group and it is commonly between the two rhyolites of that group or near their contact. The two small mapped bodies on the east side of the ridge south of Bulldog Mountain are clearly crosscutting, as is shown by their contacts where exposed in several places. The other bodies on this ridge are so poorly exposed that their relations could not be determined. They probably represent in part gently dipping dikes or sills, perhaps in part irregular intrusives. The small masses of this rock just north of Creede are somewhat better exposed and are, at least in part, crosscutting bodies. Dikelike bodies of this rock are crossed by the Nelson tunnel. The body south of Weaver has the form of a sill.

The porous character of the rock suggests a surface rock, but the texture rather suggests an intrusive. The intrusion probably took place very near the surface, and, indeed, a considerable part of the rock may have reached the surface.

PETROGRAPHY.

Megascopic features.—In luster the rocks are dull and chalky, in color they are chiefly nearly pure white with a very slight vinaceous cast, and a few grade to Quaker drab. They are commonly somewhat porous, and some carry abundant very fine pores. Much of the rock shows an imperfect fluidal banding. It carries a few thick tablets of glassy orthoclase nearly 1 centimeter across in a very fine textured to aphanitic groundmass. Dark minerals are almost entirely lacking. The rock just south of Weaver is much altered and differs somewhat from the rest of the rock. In addition to the usual large crystals of orthoclase it carries a few crystals of quartz, biotite, and a completely altered feldspar, probably a plagioclase. It has a somewhat brecciated appearance.

Microscopic features.—The thin sections show that, in addition to the large phenocrysts of orthoclase, there are a very few partly resorbed crystals of biotite and the usual accessories, apatite, zircon, and magnetite. Quartz phenocrysts and kaolinized feldspar, probably plagioclase, were found only in the specimens from the body south of Weaver. The groundmass is irregular in texture. It carries very abundant grains of quartz, of which the largest are a few tenths of a millimeter in cross section, and a few of orthoclase; these are embedded in and not sharply distinct from a very fine matrix of quartz and orthoclase. The quartz of the groundmass adjacent to the quartz grains has oriented itself with respect to these grains and is crystallographically continuous with them.

QUARTZ LATITE PORPHYRY.

OCCURRENCE.

In the drainage basin of Miners Creek and to the west there are a number of dikes, some of them several hundred feet across, of a quartz latite porphyry of rather uniform character. Most of these dikes are nearly vertical and strike a little west of north. West of the area mapped are several more dikes of the same character and striking in about the same direction, with some tendency to radiate from the hill north of MacKenzie Mountain.

The dike on the west boundary of the area is the largest of those shown, but some farther west are still larger. This dike is over 200 feet wide in places and has been followed for more than a mile. It is a part of a larger body most of which lies beyond the area mapped. The dike just west of Miners Creek is somewhat narrower and is about three-quarters of a mile long. The bodies near the fault east of Miners Creek probably represent three distinct dikes, two of them about half a mile and the other only a few hundred feet in length. The southernmost of these dikes was broken by the fault. Four small, poorly exposed bodies in the southwestern part of sec. 34, are mapped, and poor exposures of another small body were found on both sides of Rat Creek in the southern part of sec. 27. The dikes are rather resistant to weathering and commonly give prominent outcrops. Where they cut soft rocks they stand out as broken walls.

AGE.

The dikes clearly cut the rhyolites of the Alboroto group, and to the west they cut the rocks of the Piedra group, reaching as high in the section as the andesite. They are older than the last movement along the Kreutzer fault, but may possibly be younger than the main movement, as will be shown in the chapter on structure. The dike rock resembles the Fisher quartz latite both in composition and in habit, and in places north of MacKenzie Mountain there is some indication that the two grade into each other. The dikes are therefore of about the same age as this quartz latite and probably represent the channels through which it or very similar and closely related flows were extruded.

PETROGRAPHY.

Megascopic features.—In color the rocks vary somewhat but are with few exceptions gray.[40] They are dense and show phenocrysts, commonly several centimeters across, nearly equal in amount to the aphanitic groundmass. The phenocrysts are chiefly of white plagioclase in thick plates with rounded to hexagonal outline. Biotite is

---

[40] Near Ridgway's light olive-gray (23′′′′′*d*) or light mouse-gray (15′′′′′*b*).

rather abundant, and in some of the material augite and less commonly hornblende are present. Phenocrysts of orthoclase and quartz are rare.

Microscopic features.—The microscopic study shows that the plagioclase phenocrysts are not greatly zoned and are andesine-labradorite in average composition. The biotite is commonly resorbed. The accessories are apatite, zircon, and magnetite. Less than half of the rocks show remnants of augite, largely altered to calcite and chlorite; nearly all show areas of calcite and chlorite which probably represent altered augite; a few show in addition a little green hornblende. The groundmass is usually a rather coarse micrographic intergrowth of quartz and orthoclase with some small laths of plagioclase. The rocks are commonly somewhat altered, with the development of calcite and sericite from the plagioclase and of calcite and chlorite from the augite.

INTRUSIVE BASALT.

OCCURRENCE.

A single small dike of dark-colored rock was found in the Creede area. It cuts the rocks of the Piedra group as high in the section as the Windy Gulch rhyolite breccia on the slopes west of Rat Creek, in the southern part of sec. 22. It has a width of about 10 feet and strikes nearly north. It is rather poorly exposed.

PETROGRAPHY.

The fresh rock is iron-gray,[41] but on exposure it bleaches to deep olive-gray.[42] The rock is dense and shows to the naked eye a very few glassy crystals of plagioclase. With a pocket lens numerous minute laths of plagioclase can be recognized by their bright cleavage faces.

Thin sections of the rock show that it is made up in large part of small, thin laths of plagioclase, arranged nearly parallel from flow and embedded in a smaller amount of glass that is clouded with opacite and with incipient crystallization. Augite grains and prisms are rather abundant, and olivine was originally about equally abundant but is now altered to a rather strongly birefracting fibrous serpentine. The plagioclase crystals are labradorite. The usual apatite and black iron ore are present. The rock is fairly fresh except for the serpentinization of the olivine and the deposition of a little secondary calcite.

[41] Ridgway's 23′′′′′′k.
[42] Ridgway's 23′′′′′′.

HORNBLENDE-QUARTZ LATITE PORPHYRY.

A quartz latite porphyry rich in hornblende and poor in biotite and distinctly nearer the andesites than the quartz latite porphyry that is present in dikes near Sunnyside was found only in the underground workings along the Amethyst fault and is everywhere much altered. It originally contained about 30 per cent of phenocrysts as much as 3 millimeters long, chiefly of plagioclase with considerable hornblende and a little biotite.   The hornblende and biotite were much resorbed. The groundmass is made up of plagioclase laths in a very fine aggregate of quartz and orthoclase.   The hornblende is now completely altered to chlorite, carbonate, and iron oxide, and the plagioclase is also much altered and now consists of albite and oligoclase.

Chapter VIII.—QUATERNARY DEPOSITS.

SIGNIFICANCE.

A comprehensive knowledge of the topography and of the Quaternary deposits of this area is important to the geologist, as it enables him to interpret more or less imperfectly the geologically recent history of the area, such as the extent and character of the glaciation and the carving out of the present mountains and valleys. Such a knowledge should also be of considerable value to the miners and prospectors, as it aids in an understanding of enrichment, of the disappearance of some veins along the strike, such as the Amethyst vein to the north, and of other important features of the ore deposits. In the early days of the Creede camp it would have enabled the prospectors and operators to avoid the useless driving of tunnels and shafts into the terminal moraines or great landslides.

As may be seen on the geologic map of Creede and vicinity (Pl. II), Quaternary deposits cover considerable tracts in this area. On the basis of their origin and relative age they have been divided on the map into a number of units.

The earliest of these deposits, with the possible exception of some gravel deposits near Bachelor, comprise the terminal moraines of the earlier glaciers and their outwash deposits, which are probably represented by the upper terrace deposits. Next were formed the moraines of the later glaciers and their outwash benches. Nearly or quite all the landslides occurred later than the glaciation. The alluvial fans and other deposits of alluvium are in large part recent and have been enlarged and modified up to the present time.

MORAINES.

DISTRIBUTION.

Thousands of years ago (though geologically only one short period ago) the high mountains and divide north of Creede were covered by a great thickness of perennial snow. These great snow fields were the gathering grounds for many glaciers, which occupied the upper valleys of all the larger streams, such as Miners Creek, Rat Creek, and both forks of Willow Creek. The glaciers in these streams did not reach the valley of the Rio Grande but ended at an elevation of 10,000 feet. The glacier that occupied the upper basin of Rat Creek and those of the two forks of Willow Creek came into the area shown on the accompanying map (Pl. II), but that of Miners Creek ended a mile or so northwest of that area.

GLACIAL TOPOGRAPHY.

In going up any of the main streams, such as West Willow Creek, one is struck by the marked change in the character of the valleys in passing from the lower unglaciated to the upper glaciated part. Below the glaciation the valleys are all strikingly V-shaped canyons, with steep rock walls coming to a sharp angle in the creek bed. In the glaciated part the valleys open up and have almost continuous, nearly flat parks and meadows near the streams. They are U-shaped in cross section, in contrast to the V-shaped lower unglaciated parts. This contrast is shown in Plate IV (p. 3).

TWO STAGES OF GLACIATION.

Three stages of glaciation in the Quaternary period have been recognized in the San Juan Mountains.[43] Each of these stages covered a long time and in each the glaciers crept down the valley and then gradually receded. After a long interval the glaciers again increased in size and the next stage began. Only two of these stages have been recognized near Creede, and their deposits are not easily distinguished from each other.

EARLIER MORAINES.

The earlier glacier of Rat Creek probably ended northeast of MacKenzie Mountain, and the gently sloping, hummocky area just north of the hill with the 10,800-foot contour, which is about half a mile northeast of MacKenzie Mountain, represents its terminal moraine. The numerous large landslides of this area have masked the glacial material to a considerable extent. In West Willow Creek the earlier glacier reached nearly to the mouth of Nelson Creek and may have crossed the ridge into Nelson Creek. Its terminal moraine has the usual hummocky, imperfectly drained surface, but it is not so well preserved as that of the later glacier. The separation of the two moraines is only approximate. The earlier glacier reached nearly or quite to Phoenix Park in East Willow Creek, and its morainal material extends for only a short distance up the creek before being covered by the later moraine.

LATER MORAINES.

In the Creede area the later glaciers reached within a mile or so of the maximum extension of the earlier glaciers. Their terminal moraines are somewhat better preserved than those of the earlier stage and in general are but little modified and have the hummocky surface, the lack of drainage, and the numerous ponds and lakes char-

---

[43] Atwood, W. W., and Mather, K. F., The evidence of three distinct epochs in the Pleistocene history of the San Juan Mountains, Colo.: Jour. Geology, vol. 20, pp. 385–409, 1912.

acteristic of glacial moraines. In Rat Creek much of the area of the moraine is on steep hillsides, but in both forks of Willow Creek considerable areas have little relief.

CHARACTER OF MATERIAL.

Gravels make up nearly all the morainal deposits. The pebbles are of the volcanic rock from the upper part of the drainage basins, are well worn, and are in large part rather fresh. Many are flat, and some show typical glacial striations, although on the whole they are not of such a character as to preserve the markings well.

POSTGLACIAL EROSION.

Since the disappearance of the glacial ice the agents of erosion have been active but have as yet modified the canyons but little. About half a mile southeast of the northeast corner of the area included in the map East Willow Creek is in a canyon about 100 feet deep, with a nearly vertical wall of quartz latite on the east and steep slopes in the glacial débris on the west. This is an exceptionally deep postglacial canyon, and is very local and cut in comparatively soft rocks. In most places the postglacial erosion has been slight.

ECONOMIC CONSIDERATIONS.

Much prospecting has been done in the glacial material, especially along West Willow Creek. Few of the prospects have reached bedrock, as the glacial cover has a considerable thickness even where it was deposited against steep hillsides. Anyone prospecting in this material should understand that the mineral veins do not extend into the glacial deposits; that the glacial débris varies greatly in thickness and is in most places more than a thin mantle, and that it was deposited in a valley comparable in form to the present valley but somewhat deeper. Tunnels, especially in the flatter parts of the surface, will generally have to be run for considerable distances before bedrock is reached.

TERRACE GRAVELS.

In the valley of the Rio Grande to the south of the Creede area and in the lower parts of the smaller valleys are two rather prominent gravel-covered terraces respectively about 50 and 100 feet above the streams. These terraces are remnants of older valleys, somewhat higher than the present valleys and considerably broader. The upper terrace was probably formed during the earlier glacial stage by the streams that emerged from the lower ends of the glaciers as a result of the melting of the ice. The lower terrace was probably formed during the later glacial stage.

In addition to the two rather prominent and nearly continuous terraces the area contains remnants of higher gravel-covered terraces. These gravel terraces were found only on the lower parts of the streams and are in large part cut in the softer rocks. They extend up the valley of Rat Creek for some distance. The details of their distribution are shown on Plate II. As topographic features they are rather prominent, especially in the valley of the Rio Grande just south of the area included on the map. They form rather gently dipping benches on both sides of the river and have been dissected by the larger gullies. The benches are covered with gravels made up of well-rounded boulders with considerable finer material, and probably represent material carried down from the glaciers.

LANDSLIDES.

GENERAL FEATURES.

In the Creede area, as in other parts of the San Juan Mountains, landslides are abundant and some are of considerable size. Favorable conditions for their development are afforded by steep slopes in which rather thick lava flows overlie soft tuffs; slopes carved on thick flows that rest on harder rocks—for example, the lower rhyolites—are not favorable. In general, the slopes of the canyons of the four main streams are very steep, and the glaciated parts are especially favorable for landslides.

Landslides were no doubt formed before the glaciation of the region, but the greater part, at least of those shown on the map, are postglacial, and their formation has continued to the present time.

In the mapping those bodies have been considered landslides which were formed by large masses of rock moving down the slopes more or less as units. In some landslides a great mass of rock broke from the top of a mountain and moved rapidly down the slopes; in many the movement took place in stages. Some of the areas mapped as landslides represent aggregates of a number of small slides. Near the mouth of the Nelson tunnel the material is merely talus. In their surface forms landslides, like glacial moraines, are characterized by hummocks, lakelets, and imperfect drainage. However, the sags and ridges of landslides are lateral to the valleys, while those of moraines are transverse; moreover, landslides commonly show jagged cliffs, pinnacles, and similar forms.

The material that makes up a landslide is derived from the slopes above, commonly from a few bodies of rock or even a single body—the harder rock near the crest of the mountain. The fragments are characteristically angular, and some are hundreds of feet across. There is no sorting, and the arrangement is prominently chaotic. Some of the great blocks resemble cliff outcrops of rock in place except for their inconsistent structure and relations and general "jumbled up" arrangement.

DISTRIBUTION.

The distribution of landslides is shown on Plate II, and only a few special features will be discussed here. No attempt has been made to map any but the larger bodies. The large slide northwest of Bulldog Mountain was formed by the slumping of great blocks of the thick, resistant flow of tridymite latite, which is here underlain by soft beds of tuff. It is made up of exceptionally large blocks, some of which give nearly vertical cliffs of the latite, but little broken and as much as 100 feet in height. However, the chaotic arrangement and the undrained, hummocky topography are characteristically those of a landslide area.

The landslide southwest of Nelson Mountain is made up in large part of the Fisher quartz latite tuff and has some of the characteristics of a mud flow. The material moved down the slopes while soft and acted in some measure as a fluid mixture of rock and water.

The long body southwest of Mammoth Mountain might be called a rock stream. It came from the great cliffs of the Willow Creek rhyolite. The rock mass that broke from the cliffs readily crumbled into small fragments and flowed down the small gully, much as a large mass of crushed rock or coal flows down a chute of gentle incline, forming a crumpled, hummocky surface.

ECONOMIC CONSIDERATIONS.

Considerable prospecting has been done in the landslides, notably east of Deerhorn Creek. Evidently any bodies of ore found in a landslide will be small and discontinuous, and the ledges in place from which they came will be on the slopes above. If the object of the prospecting is to uncover a vein in the bedrock beneath the slide rock, an estimate of the thickness of the mantle of slide rock is important. The thickness of nearly all the landslide material indicated on the geologic map (Pl. II) is probably to be measured in tens of feet and that of much of it in hundreds of feet.

TALUS.

Accumulations of rock débris at the bases of steep slopes from the weathering of the rock above cover considerable areas and obscure the bedrock in many places. No attempt has been made to map these consistently, as they are by nature indefinite bodies. The large talus accumulations above the Nelson tunnel prevented a satisfactory interpretation of the bedrock geology and are indicated on the geologic map (Pl. II) in the same pattern as the landslides.

ALLUVIAL FANS.

Where the smaller streams and gulches reach the valleys of the larger streams at torrential stages they deposit their loads of sand

and rock, which accumulate as more or less conical or fan-shaped bodies called alluvial fans. These fans, which have been separated in a general way from the main alluvium, are found chiefly in the area of gentler topography near the valley of the Rio Grande.

ALLUVIUM.

The alluvium as here mapped includes the recent deposits of the present streams, except the alluvial fans. Most of it is ordinary stream wash of sands and gravels and is but little above the present stream channel. It is confined almost entirely to the lower courses and upper glaciated portions of the streams, as the canyons are too sharp and the stream gradients too steep for its extensive accumulation elsewhere.

The body on Rat Creek about a mile northwest of Bulldog Mountain represents the filling of a small lake formed by the large landslide from the east.

Chapter IX.—STRUCTURE.

The mountains for many miles on all sides of Creede are made up of volcanic rocks of much the same character and belonging to the same general volcanic era. Their structure is simple on the whole, and it is characterized by gentle tilting toward the Rio Grande and by a few zones or areas of block faulting. The Creede mining district embraces part of such a faulted area.

In a number of places the rocks near the faults or between two faults show considerable dips and have evidently been tilted during the faulting. In others, as at the north end of the Alpha fault, the faults appear to merge into zones of steep dip, probably associated with more or less minor faulting. Except in the immediate vicinity of the faults the rocks lie nearly flat and probably dip gently toward the south.

FAULTING.

DIFFICULTIES IN DETECTING FAULTS.

Considerable difficulty was experienced in locating and interpreting the faults in this area, chiefly on account of the great and varying thickness of nearly all the rock bodies, the lack of regular bedding planes and horizon markers, the common pinching out within a very short distance of one or more of the geologic units, the general lack of a regular and persistent succession of rocks, the marked irregularities at the bases of a large number of the rock bodies, the cover of Quaternary material on critical areas, and in places the difficulty of distinguishing between some of the rocks. The irregular character of the faulting has added to the difficulty. A glance at the geologic map (Pl. II) will amply illustrate most of these points. However, a satisfactory interpretation of nearly all the major faulting is made possible by the great relief, the comparatively good natural exposures, and the considerable amount of prospecting. Although some points, such as the northern extension of the Amethyst fault, have not been settled beyond doubt, the mapping on Plate II is believed to be essentially correct.

GENERAL CHARACTER OF FAULTING.

Most of the major faults are normal faults and strike a little west of north. The Amethyst fault and the Solomon-Ridge fault dip steeply west; the Bulldog Mountain and Alpha faults dip east. The block northeast of the Corsair mine is bounded by a nearly vertical northwest fault and by a poorly exposed southeast fault. South of the Commodore mine the Amethyst fault breaks up into a number of faults with varying strike, but their general course is considerably more to the east than that of the main fault. The great Equity fault strikes nearly east and dips very steeply north; it is the only reverse fault that has been recognized in the area mapped.

In general brecciation of the walls adjacent to the faults is not extensive, and commonly the walls are clear cut and slickensided. Not uncommonly, however, as in the Amethyst fault at the Commodore mine and in the Ridge fault near the Ridge and Solomon mine, the faults branch, forming large horses, or bodies of rock inclosed between two branches.

One of the most striking characteristics of the faults is the manner in which they die out along the strike. The great Amethyst fault, southeast of the Commodore mine, breaks up into a number of faults, some of which have throws of over 1,000 feet. North of the Park Regent mine the throw of the Amethyst fault is believed to decrease rapidly and is probably not great south of the Equity mine, but here the great Equity fault joins it, and beyond the junction it again has a great throw for some miles; farther north it has not been recognized.

The Alpha and Ridge faults pass into folds at their north extremities, probably with considerable fracturing, and are lost. Some faults, as the Equity at its west extremity, end abruptly against other faults. Sharp turns are not uncommon, as in the Amethyst fault at the Commodore mine and the Alpha fault near the Kreutzer mine.

AGE.

Nearly or quite all the faulting is believed to have taken place at about the same period in the history of the region, and it is probable that the chief displacement occurred during a rather brief geologic epoch. This epoch preceded metallization. However, minor movements have taken place since the metallization, as is indicated by some crushing and slickensiding of the vein material in nearly all the veins. The age of all the faults can not be positively determined, owing to the meager distribution of the later volcanic rocks. But the Alpha fault cuts both the Fisher quartz latite and the intrusive quartz latite porphyry, and no fault is interrupted by later igneous rocks. The Amethyst fault is known to cut the Creede formation. It is therefore probable that at least the major faulting took place after the extrusion of the youngest volcanic rocks of the area.

However, in the surrounding country there is a considerable thickness of volcanic rocks younger than any in the area here described. The Fisher quartz latite represents only the base of a considerable thickness of related flows and breccia beds, and at least one later group of rocks is known. It is not necessary to assume that the faulting is later than these younger volcanic rocks, and it may have occurred in one of the more recent periods of volcanism whose products are not preserved in the Creede area, if they were ever formed there.

The faulting took place long before the glaciation. Indeed, since the faulting erosion has removed a thickness of rock measured in thousands of feet. This is evident from an examination of the geologic map (Pl. II). Faulting that takes place rapidly and reaches the earth's surface must develop along the outcrop an escarpment in which the surface of the upthrown side of the fault is raised relative to that of the downthrown side by an amount equal to the throw of the fault. Although the Alpha, Equity, and Amethyst faults and the three faults east of North Creede all have throws of over 1,000 feet, they have no escarpments and they affect the topography only in so far as they bring together rocks of different hardness. For example, East Willow Creek crosses the extension of the Amethyst fault at a point where the throw is about 1,500 feet, yet there is no change in the gradient of the stream. To permit this condition erosion must have removed 1,500 feet more of rock on the north side of the fault than on the south side. At the time of the faulting the topography therefore must have been very different from the present topography, and the faulting must have taken place at a time before the present drainage system came into existence, or at least when it was very youthful. In this connection it must be remembered that a considerable thickness of volcanic rocks younger than any found about Creede is present in surrounding areas and was probably once present in this area.

A glance at the geologic map (Pl. II) shows that many of the faults parallel the main stream and suggests a close relation between the two. However, the faults strike a little east of south, and the natural course of the stream is directly toward the Rio Grande valley, which is southeast. There is no evidence that the faults have diverted the streams from their normal courses.

Comparatively small areas, much broken up by block faulting, chiefly of the normal type but with a few reverse faults, surrounded for considerable distances by nearly flat rocks in which notable faulting is almost entirely absent, are common in great volcanic regions, and in many such regions the geologic relations indicate that the faulting was more or less closely related to the igneous or volcanic activity. The faulting about Creede was of this character and is believed to have taken place late in the period of igneous activity and before the existing mountains and canyons of the region were developed.

MINERALIZATION ALONG FAULTS.

Nearly all the faults of the region show some mineralization, and nearly all the ore produced has come from veins along faults. The chief production has come from the Amethyst vein, which is along the major fault of the area. The ore produced at the Sunnyside camp came from the vein along the Alpha fault; the Solomon, Ridge, and Holy Moses mines are along the Solomon-Ridge fault; and the Equity mine is along the great Equity fault. Of the five major faults or fault systems, the Bulldog Mountain fault is the only one that has produced no ore.

The chief production of the region has come from the Amethyst vein, between the south end line of the Bachelor claim and the north end line of the Park Regent claim. North of the Park Regent the vein has not been located, and in the network of faults into which the Amethyst fault breaks southeast of the Commodore mine no large ore bodies have yet been discovered.

AMETHYST FAULT SYSTEM.

The Amethyst fault system is the most prominent structural feature of the region, and it gains additional importance from the fact that the productive Amethyst vein lies along it. It is here, somewhat arbitrarily, taken to embrace the main fault between the Bachelor and Park Regent mines, the network of faults extending south and southeast of the Bachelor mine, to and beyond Mammoth Mountain, and the poorly exposed system of faulting from the Captive Inca shaft to and beyond the Equity mine. Strictly the Equity fault might be considered a part of this system, but it is discussed separately.

North of the Commodore mine the Amethyst fault is a clean-cut nearly straight fault with a strike of about N. 23° W. and a dip of about 50°–70° W. At the surface the western or hanging wall is the upper member of the Creede formation in the southern part and the Windy Gulch rhyolite breccia in the northern part; only a few hundred feet below the surface, in the underground workings, the hanging wall is the Campbell Mountain rhyolite, and this continues to the Nelson tunnel level. The throw must therefore be at least 1,400 feet. In that part of the fault which is north of the branch north of the Commodore mine there are no sharp changes in direction of the fault plane, and no branching faults of any considerable size have been recognized except a possible fault of unknown throw in the hanging wall just south of the Last Chance mine. In the Last Chance and New York mines there is a great amount of fissuring and crushing of the hanging-wall rocks. Throughout this distance exposures are fairly good. In Plate IV, A (p. 3), the position of

A. FAULT EAST OF SUNNYSIDE.

B. SUNNYSIDE, COLO.

the fault is indicated by the large dumps about the shafts of the principal mines of the district.

North of the south end line of the Commodore claim the fault branches toward the south. The details of this faulting are shown on Plates I, II, and XII and figure 18. The main branch is on the west, and the combined throw is at least 1,000 feet and probably much more.

The two branches come together again near the south end of the Bachelor mine. To the southeast exposures are poor, but there must be a great fault, striking nearly southeast and well exposed at only one small outcrop about halfway between the Nelson tunnel and the portal of the adit which on the map is designated the Commodore mine. Beyond this there are no surface exposures west of West Willow Creek. Farther south and east there is a network of related faults, and no one of them alone can be considered the continuation of the Amethyst fault.

A great fault with a throw of about 1,500 feet crosses the nose of the ridge between the forks of Willow Creek. It strikes about S. 60° E. and has a steep dip. It is indicated chiefly by the manner in which the Campbell Mountain rhyolite abuts against the Willow Creek rhyolite, but it is poorly exposed in one tunnel, on the east side of the ridge. This fault continues with about the same course and crosses the ridge south of Mammoth Mountain. In some of the tunnels where it is best exposed it shows considerable brecciation in the hanging-wall rock and some mineralization. Where it separates the Campbell Mountain and Willow Creek rhyolites its position is approximately shown by the abrupt change from the cliffs of the Willow Creek rhyolite to the gentle talus-covered slopes of the Campbell Mountain rhyolite. Southwest of Mammoth Mountain, however, it joins at an acute angle a fault which lies south of it and which has a large throw in the opposite direction. East of this junction the fault probably continues to and beyond the east boundary of the area included on the map but with a throw that is probably not more than 200 feet.

The southern of these two faults can nowhere be clearly seen, but the rocks near the fault on both sides are well exposed, and its position is indicated by a change from the cliff-like outcrops of the Willow Creek rhyolite on the south to the gentler talus-covered slopes of the Campbell Mountain rhyolite on the north. It strikes about S. 70° E. and has a downthrow on the north side estimated at several hundred feet. Its dip is not known, but it probably stands nearly vertical. This and the fault to the north leave a narrow sharp-pointed wedge of the Campbell Mountain rhyolite surrounded by the Willow Creek rhyolite. This triangular block is bounded on the west by a third great fault with downthrow on the west side. This third fault strikes a little east of south. It is nowhere well exposed, but its presence east of East Willow Creek is shown by the Campbell Mountain rhyolite forming gentle slopes at the base of great cliffs of the Willow Creek rhyolite. Its position could not be accurately located, especially south of East Willow Creek, as talus covers much of this area. Where it crosses the creek it has a throw of about 1,000 feet, and beyond the point where the fault that trends S. 70° E. joins it the throw is increased to about 1,500 feet. Toward the south, after being joined by a number of small cross faults, it is covered by landslide and talus. It is believed that this fault rapidly decreases in throw toward the south, partly by tilting of the fault blocks, partly by the cross faulting, but it probably does not die out before reaching the Creede formation, although it was not located in that formation.

To the west of this north-south fault are four cross faults with throws as great as 250 feet or more, as shown on the geologic map (Pl. II). These faults were located largely by the displacement of the base of the Campbell Mountain rhyolite.

The three faults on the opposite side of Willow Creek are all of small throw. That just north of Windy Gulch, forming the south boundary of the Campbell Mountain rhyolite, is shown in a few prospect pits. It has a throw of less than a hundred feet. The other two faults are indicated chiefly by slight displacement of the intrusive rhyolite.

Just west of West Willow Creek, extending southward from the main Amethyst fault, there is probably a fault of considerable throw. This fault is indicated by the fact that on the east side of West Willow Creek the base of the Campbell Mountain rhyolite is nearly flat and at an elevation of about 9,050 feet, whereas on the west side of the creek, less than a quarter of a mile away, it is at an elevation of 9,750 feet. Moreover, in the Nelson tunnel a few hundred feet from the portal the Campbell Mountain rhyolite forms the hanging wall of the vein, although on the hill to the west its base is about 400 feet higher.

The northern extension of the Amethyst fault beyond the Park Regent mine can not be so positively determined, owing to the lack of exposures at critical places over much of the area and to some uncertainties in the interpretation of the bedrock geology west of Deerhorn Creek. From the Happy Thought shaft to Deerhorn Creek the glacial moraine covers the bedrock nearly everywhere, and for half a mile beyond the creek landslide material covers the line of the fault. Beyond the Park Regent mine there are only a few prospect shafts that show bedrock, and the interpretation of the geology of this area is based on data procured from these shafts and from a single outcrop.

A short distance north of the Park Regent shaft numerous surface prospects and the underground workings show that the fault turns rather sharply to the east and strikes nearly due north. This strike probably does not continue far, because there is no place on Nelson Mountain or to the west as far as the main branch of Deerhorn Creek where a fault of any considerable throw could pass without being recognized, and no such fault was observed. Nelson Mountain itself shows good exposures and is made up of rather regular alternating flows and tuff beds, in which a fault of even small throw could easily be recognized. To the west of Nelson Mountain, at an elevation of 11,500 feet and near the north border of the area included on the map, there is an abrupt change in the bedrock and a large area of the Equity quartz latite is exposed. This rock is not found farther southeast. It is separated from the other rocks of the Alboroto group to the southeast by a considerable landslide, and its relations here are not entirely clear. However, it is a part of a great flow, or perhaps of several great flows, which extend northward to the Equity mine and here clearly overlie the Campbell Mountain rhyolite. North and northwest of Nelson Mountain this quartz latite clearly underlies the tuffs and flows that form the upper slopes of the mountain, and if the latite is separated from the tuffs and flows by a fault on the west slope the fault can have only a small throw. The downthrown side, moreover, would be on the east instead of on the west, as in the Amethyst fault. A fault might pass into this thick quartz latite series, but no evidence of such a fault was found, and a fault with a throw comparable to that of the Amethyst fault should bring down some of the higher rocks to the west or otherwise indicate its presence.

There is some evidence as to the position of the Amethyst fault in this moraine-covered area. In the bed of West Willow Creek just west of the point where the road crosses the creek there is a fair outcrop of the andesite, and a few hundred yards to the south are some prospects that disclose the tuff and the andesite. This area is therefore similar to the isolated area of good exposures just northwest of the mouth of Deerhorn Creek and belongs in the Piedra group. The shallow shaft about 200 yards southeast of this road crossing (see Pl. II) brings up only altered rhyolite of a type that can be identified with reasonable certainty as belonging to the Willow Creek rhyolite. The main fault can therefore be placed with fair assurance between the road and this shaft.

Farther north, beyond Deerhorn Creek, the only probable position for a fault is between the rocks of the Piedra and Alboroto groups. The contact is nowhere well shown in this area, but more or less alteration and mineralization along this line indicate a fault as does also the nearly straight contact with the rocks of the Piedra group below steep slopes of rocks of the Alboroto group. The contact is not in itself conclusive evidence of a fault, as similar relations elsewhere are due to the very rugged surface over which the rocks of the Piedra group were extruded. However, a fault of considerable throw is believed to separate these two groups and is so shown on the geologic map (Pl. II).

This fault probably forks at about the crest of the ridge; one branch separates the rocks of the Piedra and Alboroto groups and the other passes into the Equity latite. This second branch is indicated largely from the relations of the rocks near the bed of West Willow Creek just below the Equity mine. The rocks of this area are all of the Alboroto group. In the creek bed there are good exposures of the Equity latite, but on the steep slopes to the east the base of this formation is about 100 feet above the creek bed. This relation continues to the Equity fault, which ends against the north-south fault. North of the Equity fault the throw of the north-south fault is much greater. This fault brings the Willow Creek rhyolite against the Equity latite. It continues just east of the creek bed for a mile or more to the north, but it probably dies out a short distance farther on in that direction, as it could not be found on the divide at the head of the creek. Its position is shown by an inconspicuous series of flats and sags along its line, which are no doubt erosional features and are due to the softness of the rhyolite to the east as compared with the latite to the west.

The displacement of the Amethyst fault or of the fault zone of West Willow Creek near Deerhorn Creek is probably not great, as the base of the Nelson Mountain quartz latite is at an elevation of about 11,600 feet on the west side of the valley and on Nelson Mountain, whereas 1¾ miles away it is at 11,900 feet, a difference of only 300 feet. The base of this flow is fairly level and regular on both sides of the creek.

EQUITY FAULT.

The Equity fault is one of the few structural features of this area that stand out clearly enough to be correctly interpreted from a hurried examination. South of the fault the Equity latite gives the rugged outcrops and broken cliffs of a hard, resistant rock. To the north along the fault line, which runs straight up the hill over very steep slopes for more than 1,000 feet, outcrops of this type give place abruptly to steep grass-covered or talus-covered slopes of the Willow Creek and Campbell Mountain rhyolites. A zone of white and iron-stained rock follows the fault plane. The fault is shown in Plate VIII, A, which is a reproduction of a photograph taken from a point on the road opposite the Equity tunnel. The fault, which is reversed, strikes a little south of east and dips steeply to the north. In the area discussed in this report it is entirely in rocks of the Alboroto group. The footwall on the south consists of the Campbell Mountain rhyolite near the creek bed but in great part of the Equity latite. The north wall consists of the Willow Creek rhyolite in the lower slopes and the Campbell Mountain rhyolite on the upper slopes. A quarter of a mile north of the place where the fault crosses the ridge between West Willow and Deerhorn creeks, and at an elevation only 100 feet or so higher, the Equity latite overlies the Campbell Mountain rhyolite. The throw of the fault is about 1,200 feet, as measured by the displacement of the base of the Equity latite.

The Equity fault ends on the west against the Amethyst fault, just east of West Willow Creek. To the east it crosses Deerhorn Creek and passes into the drainage area of East Willow Creek but with a much diminished throw.

There is considerable brecciation and mineralization along this fault, and in the Equity mine, from which the fault is named, a vein along it is worked.

FAULT BLOCK OF MAMMOTH MOUNTAIN.

On Mammoth Mountain there are three faults that may be closely related to the Amethyst system. They form a part of the boundary of the body of Campbell Mountain rhyolite. (See Pl. II.) The east-west fault on the north is probably the largest and has a throw estimated at 500 feet. It is well exposed only in the cliffs on the west slope of the mountain, where it is nearly vertical. It separates the Willow Creek rhyolite from the Campbell Mountain rhyolite and farther east from the Mammoth Mountain rhyolite.

The fault that separates the Willow Creek rhyolite on the west from the Campbell Mountain rhyolite is well exposed in the cliffs at its northern part. Its throw is at least 100 feet and is probably not much greater.

The third fault of this group separates the rhyolites of the Alboroto group from the Mammoth Mountain rhyolite and runs into the east-west fault. This fault is somewhat uncertain, as the only evidence of its presence is the form of the contact, and this may be accounted for as due to the great irregularity at the base of the Piedra group. The throw of this fault is unknown but is probably considerable.

BULLDOG MOUNTAIN FAULT.

West of Windy Gulch is a fault that strikes a little west of north and dips about 50° E. On the east slope of Bulldog Mountain it displaces the base of the andesite about 200 feet. To the south it dies out or is lost in the poorly exposed Windy Gulch rhyolite breccia; to the north it can be followed more or less continuously to the Rat Creek road. Its throw decreases toward the north, and it disappears in the tuff. There has been some prospecting along this fault, but where exposed it is not strongly mineralized.

SOLOMON-RIDGE FAULT.

The Solomon-Ridge fault is west of East Willow Creek. It has a small throw but is economically important because along it are the veins of the Solomon, Ridge, and Holy Moses mines. It strikes a few degrees west of north and dips steeply to the west. The downthrown side is on the west, and the displacement is probably only a few hundred feet. To the south it has been recognized only a few hundred yards south of the Solomon tunnel. Near the Solomon tunnel it branches, but the two branches come together again north of the Ridge mine. It was followed northward to a point nearly a mile north of the Holy Moses mine and is lost on the slopes west of Phoenix Park, in an area where exposures are very poor and interpretation of the geology somewhat uncertain. This area shows some fracturing and mineralization, which are probably related to the Ridge fault.

ALPHA FAULT.

The Alpha fault, east of Miners Creek, is characteristically crooked. Along it are the Alpha, Corsair, and Kreutzer mines, and north of the sharp turn at the Kreutzer mine it has been prospected nearly to the crest of MacKenzie Mountain.

In its southern part between the Alpha and Kreutzer mines its average strike is about N. 31° W., and it dips 54°–65° E. The hanging wall is commonly much fractured. The fault throws down the Campbell Mountain rhyolite on the east against the Willow Creek rhyolite. The displacement along the fault in this southern part is not known but is believed not to be great. A narrow much-fractured dike of quartz latite porphyry forms the southwest wall in parts of the Alpha mine; a little to the north a similar dike forms the northeast wall and is probably the same dike, indicating a considerable movement of the east wall to the north. A little farther north another dike of the same rock forms the east wall of the fault for a considerable distance. The relation of these dikes to the faulting is not entirely clear, but some, possibly all, of the faulting is later than the dikes.

A few hundred yards south of the Kreutzer mine a very poorly exposed cross fault joins the main fault from the east. Its exact relations are uncertain, as it has not perceptibly displaced the quartz latite dike, although it is believed to have crossed the dike. If the dike formed a nearly continuous thin layer on the hanging wall of the main fault, as it appears to do on the surface, a moderate displacement by a cross fault might not be perceptible in a poorly exposed area. The throw of this fault is not known, but it is believed that the north side has been dropped several hundred feet near the Alpha fault and that this displacement rapidly decreases to the east. It therefore gives the Alpha fault an increased throw toward the north.

At the Kreutzer mine the Alpha fault makes a sharp turn to the north, and beyond the mine it has an average course of a few degrees west of north; it dips steeply to the east.

It is a remarkable fact that just north of the Kreutzer mine the top of the Campbell Mountain rhyolite is dropped over 1,000 feet by this fault, yet but a mile to the north, near the crest of the MacKenzie Mountain ridge, the andesite is dropped but little. This great change in the amount of displacement along the fault is explained by the dip of the rocks lying to the east of the fault. Few of the rocks of this area are of such a character as to afford opportunity for even rough estimates of dips and strikes, but the relations are brought out clearly by the mapping of the several rock bodies. (See Pl. II.) It is evident that the flows west of the fault are nearly flat or dip gently to the south, whereas those to the east, on the ridge of MacKenzie Mountain, dip at a considerable angle to the southeast, but farther south their dip swings toward the south, and in the wedge included in the angle of the main fault and north of the cross fault (see Pl. II) they apparently dip to the southwest.

North of MacKenzie Mountain the fault passes into an area covered by Quaternary deposits and is lost. For a mile beyond, however, the rocks show considerable tilting, which is no doubt related to the faulting. This structure is best seen in the tridymite latite, in which the flow lines are normally nearly or quite flat. In this area these flow lines are especially well shown in the isolated body of tridymite latite which lies west of the earlier glacial moraine. (See Pl. II.) This entire body dips to the east at 35°–40° on the upper part of the slope and at somewhat smaller angles on the lower part. Indeed, this slope is nearly a dip slope and is probably at about the top of the tridymite latite, as the andesite appears on the northwest corner of this outcrop with its base at an elevation of 11,250 feet, and less than half a mile to the east it is at an elevation of 10,450 feet. The Fisher quartz latite is probably also affected by the structure above described. Its base is nearly everywhere covered by landslide material, but over nearly the whole of this area the top of the landslide laps but little upon the latite, as is indicated by the exposure of the glassy base of the latite at numerous places. No faulting was observed in this area of steep dip, but minor faulting is probable; indeed, it is difficult to conceive how, without fracturing, these sharp changes in structure could take place in brittle rock which was under only a moderate load.

This zone of steep dips has not been recognized as far north as the later glacial moraines. It is a narrow strip, less than half a mile wide and about 2 miles long.

About a mile or less to the east of this zone, on the east side of Rat Creek, is an area of rocks which probably have considerable dips to the west, although the structure is not clearly shown and is indicated chiefly by the form of contacts, which are nowhere regular, and by the great vertical extent covered by the Windy Gulch rhyolite breccia. These two belts would make a canoe-shaped syncline in the Rat Creek basin, with the point of the canoe at about the south end of the upper glacial moraine. The continuation of this zone of complex structure to the north may account for the irregularities in the relations of the andesite and tridymite latite on Rat Creek, just south of the northwest corner of the area included on the map. Exposures are poor in this area, and no satisfactory interpretation of the structure could be made. To the south this syncline probably extends to the fault east of the Corsair mine. This small structural basin is bounded to a considerable extent by faults of large throw and is probably associated with much minor faulting.

STRUCTURE NORTHEAST OF SUNNYSIDE.

Northeast of Sunnyside, as shown by the geologic map (Pl. II), there is an area of block faulting, with tilted blocks. Poor exposures, the irregular erosional surfaces that separate the formations, the decomposition of the rocks, and their lack of development in their most characteristic forms introduce some uncertainty into the mapping and interpretation of the structure.

The rocks of this area are the Willow Creek and Campbell Mountain rhyolites, the intrusive rhyolite, the hornblende-quartz latite, the Windy Gulch rhyolite breccia, the tridymite latite, and the several members of the Creede formation. Here the Campbell Mountain rhyolite resembles both the underlying Willow Creek rhyolite and the overlying Windy Gulch rhyolite breccia, and separation is difficult. The hornblende-quartz latite is poorly exposed and is not altogether characteristic, but the tridymite latite is fairly characteristic, and the rocks of the Creede formation, except for subordinate tuff in the hornblende-quartz latite, are easily distinguished.

A nearly vertical fault, which strikes about N. 40° E., forms the northwest boundary of this area. On the west side of Rat Creek it is poorly exposed and is somewhat uncertain, but from Rat Creek toward the northeast it is easily followed. Just east of Rat Creek its line is marked for several hundred yeards by low cliffs of the Willow Creek rhyolite facing to the southeast (Pl. XI, A). The lower slopes only a few feet northeast of these cliffs show fair exposures of the tridymite latite, which is here soft and weathers into low pinnacles. The flow banding of this rock is cut off sharply by the cliff of rhyolite, and prospects along the fault expose the contact at a number of places. There is little mineralization. The throw of the fault can not be estimated, but it is considerable near Rat Creek, and it is believed to decrease rapidly to the northeast, as the latite dips to the southwest. No trace of this fault was seen south of the Corsair mine, and it probably goes under the alluvium and is lost in the alluvium and tuff. \ It may join the Alpha fault, and if so it would increase the throw of that fault to the south. Toward the northeast it can be followed nearly to the crest of the ridge, and it must within a short distance die out or join a fault running southeast.

The southeast contact of this narrow strip of tridymite latite may also be a fault. East of Rat Creek the nearly straight contact cutting across the surface features suggests a fault, although in other places normal contacts at the base of the lake beds are of this character. The contact is indicated as normal on the map, for the facts can be interpreted as well without assuming a fault to be present.

Five other faults, all of small throw, have been mapped in this area. They have throws of 50 to a few hundred feet and probably continue for no great distance along their strikes. The east-west fault on the north boundary of the tridymite latite strip is nowhere exposed, but its presence is reasonably certain from the form of this contact. The way in which the Creede formation is displaced by the easterly extension of this fault indicates that the north block moved to the west. The northeast boundary of the tridymite latite body also is believed to be a fault, on account of the form of the contact. This fault probably continues to the northwest and joins the main fault of this block. The two small faults that cut the hornblende-quartz latite are well exposed in prospect pits. The supposed fault bounding this latite on the northeast is nowhere exposed, and this may be a normal contact.

MINOR FAULTS.

In addition to the faults already described there are a large number of faults and slips of uncertain but probably small throw. In some the throw may be a hundred feet or even more, and they could be consistently mapped in an area of comparatively regularly bedded rock but not without great difficulty and uncertainty in this area. Only a few of these minor faults are indicated on the geologic map.

Chapter X.—ORE DEPOSITS.

GENERAL CHARACTER.

The ore deposits of the Creede district are silver-lead fissure veins in rhyolite and fractured zones of silver ore in shattered rhyolite. The total production, except a small amount, has been obtained from the silver-lead fissure veins. These veins occupy strong fault fissures and and in the main are extensive both vertically and along the strike. They include the Amethyst, the Solomon-Holy Moses, the Alpha-Corsair, the Mammoth, and several smaller lodes. All these veins strike in the northwest quadrant, and the majority dip west or southwest.

Faulting has taken place on a large scale, as is indicated by slickensided surfaces with abundant movement striae and at many places by a lack of correspondence of the rocks on the two sides of veins. The principal veins fill fissures along normal faults, and at some places, particularly in the hanging-wall blocks, there are subordinate fissures which join the principal faults in depth. Such relations indicate that the hanging wall of a fault was shattered as it was drawn downward by gravity along the footwall.

Some of the veins have been opened by movement since the ore was deposited. The results of such movement in the Amethyst vein are very pronounced. The ore itself is crossed by striated slickensided planes, and locally the vein quartz with associated sulphides forms a friction breccia. The ore minerals include zinc blende, argentiferous galena, gold, pyrite, chalcopyrite, and their alteration products. The gangue minerals include quartz, much of it amethystine, with chlorite, barite, and fluorite. The several veins show considerable differences mineralogically. Hydrothermal metamorphism is not so pronounced over wide areas as in many other mineralized regions, but along the most productive portions of veins considerable alteration has taken place. It is attended by silicification and the development of some sericite. Locally the gangue contains much thuringite, an iron-rich chlorite, and near the Amethyst vein adularia has been noted in veinlets cutting rhyolite. Ribbon quartz and symmetrically banded crusts are common, indicating deposition locally in open spaces. For reasons that are mentioned on page 120, it is thought that these veins have been deposited by ascending thermal waters. As they cut rocks that are regarded as of Miocene age, the deposits are Miocene or later.

In some of the deposits enrichment is pronounced. The rich secondary ores extend downward to great depths, owing to the high relief of the area and consequent ample head of the solutions and to the open character of the veins, all of which facilitate a rapid downward circulation.

The fractured zones of silver ore in shattered rhyolite include the deposits of the Mollie S. and Monte Carlo mines. The fractures and joint planes of the rhyolite are filled with thin veinlets of green chrysoprase and other green copper minerals and locally carry very high percentages of silver. Argentite, cerargyrite, and native silver are plastered on the walls of the thin, narrow cracks. Iron sulphides are not abundant. The rhyolite along the veinlets is apparently fresh and not greatly affected by hydrothermal metamorphism. Deeper exploration has not exposed corresponding bodies of sulphide ores, and it is possible that the rich ores of this class are genetically related to the present topographic surface.

MINERALS OF THE DEPOSITS.

Quartz.—It is shown by the numerous analyses of the ore shipments that nearly all the ores carry from 60 to 80 per cent of silica. Much of this, however, is in the rhyolite that is included in the ore. Vein quartz is the most abundant mineral, although it is less conspicuous in some ores than barite and thuringite. Barite is more prominent in the upper portions of the lodes and thuringite in the lower portions, but quartz is abundant throughout the deposits.

Of the many varieties of quartz which are present, four are particularly noteworthy. One of these is a dark-gray quartz, the color of which is due to finely disseminated galena, pyrite, and sphalerite. As a rule such material is ore. A second variety of quartz is milky white or slightly turbid, and generally it is barren. Another variety is of amethystine color, and the Amethyst, which is the principal lode of the district, takes its name from this mineral. Amethystine quartz is found in many of the deposits. It is especially abundant in the Amethyst vein and has been noted in the Mammoth and Overholt and very sparingly in the Corsair-Alpha. It is practically unknown in the Solomon, and none was noted in the Mollie S., Monte Carlo, and Monon mines. Amethystine quartz occurs interbanded with the sulphides of the primary ores, and some ore of this character is of good grade. Much of the amethystine quartz, however, especially that in the Amethyst and Last Chance mines, is apparently later than the first period of ore deposition. It forms crustified veins several inches thick which cut earlier ore, and at some places fragments of metallic sulphide ore approximately an inch in dimensions are surrounded by barren crustified amethystine quartz. The coloring of amethystine quartz is commonly supposed to be due to the presence of a small amount of manganese. A test made by George Steiger in the laboratory of the Geological Survey of amethystine quartz from the Amethyst vein showed that it contained only 0.0005 per cent of MnO. This is considerably less than is contained in the rhyolites that form the walls of the vein. A fourth variety of the quartz has apparently been deposited by descending waters, for it forms banded crusts that alternate with chalcedony and native silver. In ores of the Monte Carlo mine quartz crystals are deposited on siderite.

Barite.—Barite appears in considerable amounts at the outcrops of nearly all the lodes near Creede. It has been noted wherever the Amethyst vein is exposed at the surface, and it extends in abundance to the deepest levels of the Bachelor and Commodore mines. Analyses of train loads of ore from these veins show 12 to 16 per cent of $BaSO_4$, and some shipments carry over 20 per cent. Barite is not abundant in the deeper levels of the Last Chance and Amethyst mines, and it is rarer still in the deeper levels of the Happy Thought and Park Regent mines. It is common in veinlets or as a cement in the breccias at a distance from the veins, especially about Windy Gulch. .

Some of the barite, especially that in the lower levels, is pure white, but some near the outcrop is stained pink by iron oxide. On levels 5 and 6 of the Amethyst mine, about 250 feet north of the main shaft, limonite ore carries crystals of barite in numerous fracture planes. Broken pieces of the altered ore are coated with thinly spaced barite crystals, some of which are more than an inch long. In the levels below level 6 barite is less abundant. Barite has been noted also in ores of the Mammoth, Exchequer, Mustang, Equity, Kreutzer, and other mines. None was noted in the Mollie S. and Monte Carlo, and only a little in the Alpha-Corsair lode. It is very sparingly present in the upper levels of the Solomon-Holy Moses vein but was not noted on the adit level of the Solomon mine, driven on that vein.

The barite is believed to be both primary and secondary in origin. Analyses of rhyolite near the Amethyst vein show 0.49 per cent of BaO. On weathering this could be dissolved as carbonate. In the lodes, where the mineral waters carry sulphates, the insoluble barium sulphate would be deposited.

Thuringite.—Thuringite, an iron-rich chlorite, is one of the most abundant gangue minerals in the Amethyst and Solomon lodes. It is olive-green in color and occurs as small, thin, closely packed plates and fibers, forming a green putty-like material which in general carries small disseminated masses of sphalerite and galena. It oxidizes readily and was not noted on outcrops or in the oxidized zones. Little or no thuringite was noted at points as near as 500 feet to the present surface. Flakes of red mica-like particles in the lower levels of the Amethyst vein are thuringite altering to hematite or turgite. Some of the limonite and manganese oxides in the higher levels are doubtless formed by the oxidation of thuringite. An analysis by J. G. Fairchild of the chloritic material consisting of thuringite with a very little quartz, taken from the north end of level 12, Amethyst mine, is stated below (No. 1). For comparison, analyses of thuringite from Harpers Ferry (No. 2) and Arkansas (No. 3) are also given. The three analyses represent material that is very much alike. The Creede mineral (No. 1) is lower in FeO and correspondingly high in MgO and MnO.

Analyses of thuringite.

	1	2	3		1	2	3
SiO_2	24.34	23.58	23.70	H_2O-	0.35		
Al_2O_3	16.46	16.85	16.54	H_2O+	9.19	10.45	10.90
Fe_2O_3	12.04	14.33	12.13	TiO_2	Trace.		
FeO	28.89	33.20	33.14	CO_2	Trace.		
MgO	5.41	1.52	1.85	P_2O_5	Trace.		
CaO	None.			S	Trace.		
Na_2O	.37	.46	.32	MnO	2.75	.09	1.16
K_2O	Trace.						
					99.80	100.48	99.74

An optical study of the thuringite from various parts of the Creede district showed that it varies somewhat in its optical properties, and it probably varies also in chemical composition. The material analyzed is in very minute interwoven fibers with positive elongation and optically negative character. The axial angle is moderate. It is pleochroic and dark green parallel to the fibers (β and γ) and nearly colorless normal thereto (α). The mean index of refraction is about 1.637 ± 0.005, and the birefringence is about 0.01. The fibers are too small for a satisfactory optical study. A specimen of the mineral from the Park Regent mine showed essentially the same optical properties, but the indices of refraction are slightly higher ($\eta = 1.643$).

A specimen from the Ridge mine is considerably paler in color, is less strongly pleochroic, is a little more coarsely crystalline, and has lower indices of refraction but is otherwise similar. It is optically negative and has a moderate axial angle. X is normal to the fibers and plates and is very pale green; Z and Y are a somewhat darker olive-green. $\alpha = 1.595 \pm 0.005$. $\gamma = 1.605 \pm 0.005$. Another specimen from the Ridge mine is in very minute fibers and has a mean index of refraction of about 1.585 ± 0.005. A specimen from the Solomon mine is similar but has a mean index of refraction of 1.617 ± 0.005. These data indicate a moderate variation in composition. The Ridge-Solomon mineral, with its lower index of refraction, is probably higher in Al_2O_3 and MgO and lower in Fe_2O_3 and FeO. This thuringite approaches aphrosiderite or delessite in composition. The analysis of the gouge from the Solomon mine (analysis 4, p. 119), which was made up largely of chlorite but contained some quartz and chalky, decomposed rhyolite, with here and there a flake of sericite, confirms this conclusion.

Adularia.—Thin veinlets composed of adularia, vein orthoclase, and quartz cut the rhyolite wall rock of the Amethyst lode. These veinlets are of milk-white color. As some of them inclose fragments of altered rhyolite, it is believed that the deposition of adularia possibly attended at least one phase of the mineralization of the area. The great bulk of the ore, however, is free from adularia, which was sought in many samples of the normal vein matter but not found. The orthoclase associated with the chlorite ores seems to be the original orthoclase of the rhyolite. Small veinlets of adularia are found in the rhyolite that appears not to have been affected by metalliferous solutions.

Rhodochrosite.—Rhodochrosite occurs very rarely in the ore of the Amethyst vein. A small cluster of crystals lining a vug in the vein was noted on the Nelson tunnel level in the Bachelor mine near the point where the Bachelor vein joins the Amethyst fault.

Fluorite.—A little white fluorite was noted on the dump of the Holy Moses mine in material that was evidently taken from the upper levels of the Solomon-Holy Moses vein. Fluorite was noted also at the Solomon mill in the ore from the Solomon mine. No fluorite was noted in the Amethyst lode. In a vein at Wagonwheel Gap, 12 miles southeast of Creede, fluorite is abundant and has been mined on a commercial scale.

Galena.—Galena is the most valuable ore of lead and probably the only primary lead mineral in the Creede deposits. Much of the galena shows conspicuously the characteristic perfect cubic cleavage. Some fine-grained varieties in which the cleavage is less evident are termed steel galena by the miners. All the lead concentrates carry silver. In the lower levels of the Amethyst and Park Regent mines blocks of galena several feet square and 6 inches thick have been mined. Most of the galena occurs, however, as thin bands alternating with sphalerite and quartz, or as small masses disseminated in thuringite gangue. Paper-thin veinlets of galena cut the rhyolite that forms the walls along the lodes; at some places small specks of galena are embedded in the altered wall rock. Galena alters to anglesite, cerusite, and pyromorphite.

Pyrite.—Pyrite is present in the sulphide ores of all the lodes. It has been deposited in open spaces and as small crystals and masses in the wall rock. It is not abundant in most of the ores. Analyses of many shipments show less than 5 per cent of iron.

Sphalerite.—Sphalerite is one of the most abundant minerals in unoxidized ores of the Amethyst and Solomon-Holy Moses veins, and it has been noted in ores of the Alpha-Corsair vein. In the lower levels of the Amethyst lode it is intergrown with galena, pyrite, quartz, and other minerals. Some of the ore consists of alternating bands of sphalerite and galena. Most of the sphalerite in the district is yellow or resinous in color, though dark varieties have been noted. Sphalerite is more abundant in the Amethyst and Solomon veins than in the Corsair-Alpha vein. In the Solomon vein in the lower levels sphalerite is one of the principal ore minerals. In much of the ore it occurs as small masses disseminated through the gangue.

Chalcopyrite.—Chalcopyrite has been noted in small quantities in the ores of the Amethyst, Solomon-Holy Moses, and Corsair-Alpha lodes. In the sulphide ores of this district that have been examined it constitutes only a small fraction of 1 per cent.

Gold.—Gold as a primary mineral is probably included in pyrite, galena, sphalerite, and other minerals in the Creede district. As a primary mineral it is presumably very finely divided, for visible particles have not been identified in surroundings that suggest a primary origin. Some of the lead concentrates are rich in gold. In these concentrates, however, much of the gold is not intergrown with galena, but in the method of concentration the gold and galena come from the tables together. The richest gold ore consists of gold in a gangue of manganese oxide in veinlets cutting the older sulphides. Ore with such veinlets may carry as much as 1 or 2 ounces of gold to the ton, whereas the older primary ore carries as a rule about 0.1 ounce or less. A number of samples of manganiferous ore were panned to ascertain the character and association of the gold. In nearly all these the galena and lead sulphate remained with gold in the bottom of the pan. In washing the ore in a horn the gold might not be discovered, but in the ordinary gold pan with a sharp angle between sides and bottom the gold may easily be separated from manganese and lead minerals. As a rule the gold is dark yellow and apparently has a high degree of fineness. The association of gold and manganese oxide in cracks in the older ore is very clearly shown in the Happy Thought and Amethyst mines. This association is said to be common, but no clearly defined examples were noted in other deposits.

Native silver.—Native silver is an abundant constituent of the ores of the Amethyst vein and was found in the upper levels of the Solomon and Holy Moses mines and in the Corsair and Alpha veins. It has been identified also in the ore of the Monte Carlo and Mollie S. mines. In the Amethyst lode it is said to have been especially abundant in the ores of the upper levels of the Last Chance, New York, Commodore, and Bachelor mines but is only sparingly present in the Happy Thought and mines farther north on the Amethyst lode. In general it forms thin sheets in the siliceous sulphide ore or is plastered on fracture surfaces. In cabinet specimens nests of thin, closely packed wires of native silver fill little vugs in dark quartz that is stained with sulphides.

In the Commodore mine native silver is said to have been abundant in the "Wire Silver stope," which is about 1,100 feet below the surface. As native silver has not been found on the Nelson tunnel level of the Amethyst lode, nor as deep as 1,100 feet below the surface at any place on this lode except in the Commodore mine, it is believed to be of secondary origin. It was deposited by descending sulphate solutions in fractures in the earlier ore. In some of the rich ore native silver is beautifully crustified with red jasper, milky chalcedony, and thin bands of comb quartz. This silver must have been deposited with silica in open spaces. A polished surface of ore consisting of chalcedony, quartz, and native silver is shown in figure 11 (p. 117).

Argentite.—Argentite was probably a common mineral in the richer ores. It is said to have been a constituent of the richer ores of the Amethyst lode but was not noted in the Amethyst mine in the course of this investigation. It was seen in ores from the dump of the Mollie S. mine and is said to be present in the Alpha-Corsair ores. Possibly it is intergrown with lead sulphide in argentiferous galena.

Stephanite.—Stephanite is said to have been found in ores of the Alpha-Corsair lode, and a trace of antimony was noted in ore from the Corsair dump. An antimony mineral is reported from ores of the Overholt and Bachelor mines. No crystals of these minerals have been noted by the writers either in the mines or in cabinet specimens collected when the richer secondary ores were being exploited.

Marcasite.—Marcasite is developed in the superficial ores in some of the deposits but is not abundant. Masses of considerable size were noted on the dump of the Delaware prospect near Sunnyside. There the marcasite crusts over fragments of rhyolite and apparently belongs to a late stage of deposition.

Anglesite.—Lead sulphate is abundant in the oxidized ore, where it is an oxidation product of galena. It occurs as beautiful glassy crystals in cavities in galena and as massive bodies in an oxidized siliceous gangue. Many massive bodies of heavy white ore that are generally considered to be cerusite or lead carbonate have proved to be anglesite. It is doubtless the more common alteration product of lead. One specimen from the Happy Thought mine consists of massive glassy lead sulphate surrounding a cavity filled with anglesite crystals. In the upper levels, under oxidizing conditions, nodules of galena are surrounded by concentric layers of lead sulphate. One of these is shown in figure 14 (p. 124). In the Park Regent mine, on the Nelson tunnel level, more than 1,000 feet below the surface, thin veinlets of anglesite have formed in fractured galena.

Cerusite.—Lead carbonate, or cerusite, is developed in the oxidized ore of the Amethyst, Holy Moses, and other mines and is said to have occurred in considerable amounts in and near the apex of the Amethyst lode on the Happy Thought claim. It was mined near the surface on the Solomon, Ethel, and Holy Moses claims. Masses of earthy carbonate 2 or 3 inches in diameter were found on the Carbonate claim. The outer shells of nodules of galena altering to anglesite carry a little cerusite, which apparently is formed after anglesite. From the data available cerusite appears to have a narrower vertical range than anglesite, which is more abundant in depth.

Limonite.—Limonite occurs at the outcrops of all the deposits but as a rule is not abundant. In the oxidized zone, especially in the Amethyst and Last Chance mines at the north end of the Amethyst lode, limonitic ores are comparatively abundant. In view of the fact that the primary ore carries only a few per cent of pyrite or other iron sulphide, it appears probable that some of the limonite, perhaps a considerable portion, has formed from the oxidation of the thuringite.

Jarosite.—Jarosite was identified in oxidized material taken near the outcrop of the Amethyst lode at the Last Chance mine.

Calcite.—In the Amethyst and Solomon-Holy Moses lodes, small crystals of calcite in altered ore and thin veinlets in the wall rock have been noted, but these are exceptional, at least in the little-altered ore of the lower levels. Calcite is abundant in the Monon mine and in the travertine deposits south of Creede.

Siderite.—Siderite was not noted in the Amethyst and Solomon veins. It was observed on fracture planes in shattered rhyolite on the Monte Carlo lode. In some of the ores of this mine siderite is incrusted with small crystals of white quartz.

Malachite.—A few small masses of malachite have been noted in ores of the Amethyst, Alpha-Corsair, and several other deposits.

Chrysocolla.—Chrysocolla was noted in the oxidized ores of the Mollie S. and Monte Carlo mines and is probably present in oxidized ores of other deposits.

Kaolin.—Kaolin is abundant, especially in the oxidized zones of the deposits. It forms as an alteration product from rhyolite and is probably the most abundant alteration product of the gangue. It is associated with cerusite in coatings over nodules of galena which has been altered to anglesite. In this association it has been deposited either from solution or from suspension by downward-moving waters.

Cerargyrite.—Cerargyrite was noted in the oxidized ores of the Mollie S. mine and is said to have been present in the superficial zone of the Amethyst lode.

Chrysoprase.—Chrysoprase or apple-green chalcedony is abundant in the ores of the Mollie S. and Monte Carlo mines. Some of the darker specimens, when polished, make beautiful gems and have been mistaken for turquoise.   According to Dana,[44] the coloring of chrysoprase is due to the presence of nickel oxide.   That of the mineral from the Mollie S. mine appears to be due to a salt of copper.

Chalcedony.—Chalcedonic quartz is a fairly common constituent of the Creede ores.   It replaces the rhyolite along the lodes and is formed by processes of superficial alteration in the veins.   Some of it has been deposited in open spaces.   Red jasper is crustified with quartz and native silver in ores from the Amethyst lode, the silica and native silver having been deposited together in open spaces by descending waters.

Wad.—Black manganese dioxide is a fairly abundant constituent of the oxidized ores of the Amethyst lode.   The oxidized ores of the Monte Carlo and Mollie S. mines carry beautiful dendritic manganese flowers on fractured surfaces.   No crystallized manganese oxides have been identified.   In the Amethyst lode probably most of the manganese has come from the decomposition of thuringite, which, as shown by analyses, carries 2.75 per cent of MnO.   Rhodochrosite has been noted but is exceedingly rare; rhodonite has not been observed.   Rich gold ores in the Happy Thought and Amethyst mines are all associated with manganese oxides.

Goslarite.—Hydrated zinc sulphate, or goslarite, is present in many old drifts.   Commonly it forms nests of heavy, hairlike material on the walls and floors of abandoned workings.   At some places—for example, on the sixth level of the Amethyst mine—nests a foot in diameter may be found.   It has evidently been precipitated from water trickling downward through the sulphide ore, doubtless by a process of drying in the presence of air.   Some goslarite is forming now.

Gypsum.—Gypsum occurs sparingly in the oxidized ores of the Amethyst vein and has been noted in the ore of the Monte Carlo and other deposits.   It is, however, comparatively rare in veins of this district.   It forms a large part of one of the travertine deposits southwest of Creede.

Chalcanthite.—Copper sulphate (chalcanthite) is found in small quantities in the upper part of the Amethyst vein and in the Corsair-Alpha vein at Sunnyside.

Melanterite.—Melanterite is common on walls in workings in the oxidized zone.

Smithsonite.—A little smithsonite is present in the superficial ores of the Amethyst vein.   In view of the abundance of sphalerite in the Creede ores and of carbonates in the mine waters, it is surprising that so little smithsonite has been deposited.

---

[44] System of mineralogy, 6th ed., p. 188, 1901.

Pyromorphite.—Some beautiful yellow and brown crystals of pyromorphite were observed in a private mineral collection at Creede. These crystals had been taken from the upper levels of the Amethyst vein in the earlier days of its development. They evidently had formed along the sides of a cavity by a process of alteration. Without doubt the lead in the pyromorphite had been deposited directly from solution.

Hematite.—The ferruginous oxidized material of the upper levels of the lodes is in general brown or canary-yellow; locally it is red. Some of this red material is probably hematite, but it is difficult to distinguish between hematite and the hydrated iron oxide, turgite. No crystals of hematite have been observed.

THE FRACTURES.

GENERAL CHARACTER.

All the workable deposits are in or along fissures, and nearly all are along faults of considerable throw. Most of the fault systems show mineralization in places; none of them, however, are mineralized throughout their courses.

The displacement of the Amethyst fault is probably 1,500 feet or more. The displacement of the Mammoth fault is about 1,000 feet. The displacement of the Solomon-Holy Moses fault is small. All the faults that carry ore are normal faults except one at the Equity mine, which is a reverse fault. The occurrence of the ore deposit of the Equity mine is noteworthy;

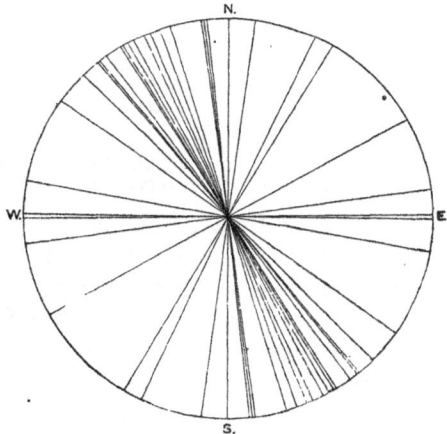

FIGURE 3.—Sketch showing strikes of principal lodes of Creede district, plotted through a common center.

examples of ore deposits in reverse faults, especially in rocks as young as the Tertiary, are rare.

In figure 3 the strikes of the veins of the district are plotted through a common center. The veins strike in every quadrant of the circle, but most of them strike in the north half of the northwest quadrant, between north and N. 40° W. The Solomon-Ridge vein strikes about N. 10° W., the Amethyst N. 23° W., and the Alpha-Corsair about N. 32° W. These three veins with their associated subordinate fissures have produced over 99 per cent of the ore of the district. About half of the less productive veins strike between N. 7° W. and N. 32° W., or in about one-seventh of a semicircle. Most of the veins, as shown by figure 4, dip at high angles, from 50° to 90°.   Not one as flat as 45° has been discovered.   The Amethyst, Mammoth, Solomon-Ridge, and many minor veins dip toward the west.   The Alpha-Corsair and three minor veins dip east.

At many places the lodes are sheeted zones; instead of a single fracture filled with ore, several closely spaced planes of movement are mineralized.   Parallel veins occur in the footwall of the Amethyst fault in the Bachelor mine and in the hanging wall in the Commodore and Last Chance mines.   The sheeting on the Corsair-Alpha lode is shown by figure 5.

In many metalliferous districts the ore deposits do not follow the conspicuous faults or those with the largest throw.   Thus in the Coeur d'Alene region of Idaho and the Philipsburg region of Montana there are many large faults, but only a few of the more productive veins are related to faults of determinable throw, and these are

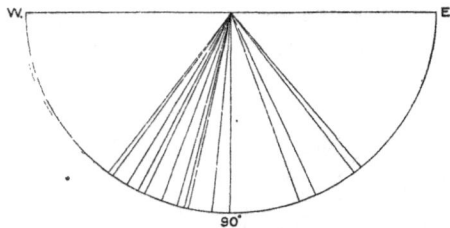

FIGURE 4.—Sketch showing dips of principal lodes of Creede district, plotted through a common center.

related to subordinate faults. In many other regions that contain numerous faults the ores are mainly in subordinate fractures or in faults of slight throw.   As has been pointed out by Ransome,[45] clay gouge is probably developed by movement, and in some deposits this retards circulation.   In Butte, Mont.,[46] there is clear evidence that the circulation of the solutions has been controlled locally in the fault fissures by the putty-like material that has been produced by movement.

There are a few districts in the United States, however, which contain deposits that have formed along faults of great throw. Among these deposits are the Comstock lode, some of the deposits of Bullfrog, Nev., the deposits in the fault fissures at Butte, Mont., some subordinate deposits at Philipsburg, Mont., and some in the Silverton quadrangle, Colo.   In the Creede district all the most productive lodes are along faults, and some of these faults have great throw.   The throw of the Amethyst fault is 1,500 feet or more, that of the Corsair-Alpha is several hundred feet, and that of the Solomon-Ridge is probably more than 100 feet.   The Mammoth fault has a throw of 800 feet or more and the Equity 1,200 feet.   It is believed, however, that the fault fissures were filled soon after they were formed.   The ores they carry are of comparatively late age,

    45 Ransome, F. L., The relation between certain ore-bearing veins and gouge-filled fissures: Econ. Geology, vol. 3, p. 331, 1908.
    46 Sales, Reno, The localization of values in ore bodies and the occurrence of shoots in metalliferous deposits: Econ. Geology, vol. 3, p. 326, 1908.

and presumably the faulting took place relatively near the surface and not under a very great load. In the Amethyst lode the larger proportion of ore is in the fault fissure itself, but much ore occurs in minor fractures in the hanging wall of the fault. In the Bachelor mine and at the south end of the Commodore mine a fissure in the foot-wall of the fault has supplied most of the ore.

The structure of the deposits and the fractures they occupy are discussed in Chapter XI in the descriptions of the mines. Here are mentioned only the more prominent features of the productive lodes.

The Amethyst vein is best exposed on the Nelson adit (Pl. XII). About 600 feet from the portal this adit encounters a dike of horn-blende-quartz latite porphyry, through which it is driven for 160 feet. This dike follows the Amethyst fault. The rock east of it is the Willow Creek rhyolite and that west of it the Campbell Mountain rhyolite. The porphyry is crushed and altered and contains some iron-stained fractures but no work-able ore has been encountered in it. About 1,400 feet from the portal the adit is turned N. 18° W., and 525 feet beyond this turn it encounters the Amethyst fault. North of this point for about 1,000 feet the tunnel follows a vein in the foot-

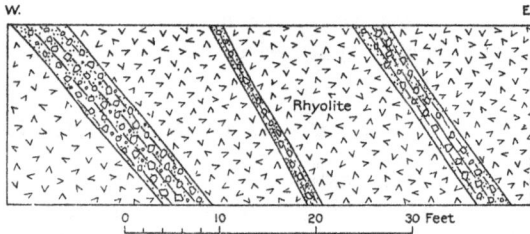

FIGURE 5.—Sketch showing sheeted zone of Alpha-Corsair lode, about 7 feet above the upper Alpha tunnel and near the Alpha boundary line. Dark areas are crushed rhyolite and vein matter.

wall of the fault, but it crosses the fault again about 40 feet farther north. From that point for about 2.000 feet the tunnel is driven in the Campbell Mountain rhyolite. For the next 560 feet it is driven in the footwall Willow Creek rhyolite, and then it crosses the fault again. At this crossing, which is about 600 feet south of the bottom of the Last Chance shaft, both walls of the fault are the Willow Creek rhyolite, but at a point 400 feet south of the Last Chance shaft the Campbell Mountain rhyolite appears on both sides of the crosscut, which is here in the hanging wall of the Amethyst fault. Evidently at this point the crosscut is near the top of the Willow Creek rhyolite, which is overlain by the Campbell Mountain rhyolite; northward from this point to the face of the adit the Campbell Mountain is the only rhyolite on the hanging wall of the fault. A hornblende-quartz latite porphyry is exposed in the hanging wall of the fault, in the crosscut to the Last Chance shaft, and in the footwall of the fault 600 feet north of the Happy Thought shaft.

Figure 30 (p. 178) is a plan of the Solomon adit on the Solomon-Ridge-Holy Moses lode. On this adit the vein system is developed for nearly 3,200 feet along the strike.   The country rock is everywhere the Willow Creek rhyolite.   The adit encounters the vein 420 feet from the portal.   At 75 feet north of this point the vein splits and the Ethel vein makes off in the footwall.   The Solomon and Ethel veins probably join again about 1,950 feet north of the split, between chutes 9 and 10.   Both veins dip west.   In the Ridge mine the east vein is called the Ridge vein and the west vein the Mexico.   The Holy Moses vein, which farther north is developed nearer the surface, is doubtless the vein that is followed in the Solomon adit northward from the intersection of the Solomon and Ethel veins.

The Alpha-Corsair lode occupies a fissure in the Willow Creek rhyolite.   Here and there on the surface and also in the Corsair adit it follows a porphyry dike.   The dike rock is highly crushed and altered and is evidently older than the fissures occupied by the ore. The porphyry appears to have had no favorable influence on ore deposition; the richest ore was found between rhyolite walls a few feet below the porphyry dike.   Pronounced splitting and sheeting are noteworthy features of the lode south of the inclined winze on the Corsair tunnel.

NATURE OF FAULT MOVEMENTS.

It is generally assumed, with good reason, that in a region of normal faulting the hanging walls of the faults move downward.   In some regions, however, the movement is not directly downward, but there is a lateral element of movement also.   In the Ridge mine, on the Solomon-Holy Moses lode, about 250 feet north of the blind shaft, striae on the vein plunge 75° S.   In the Solomon mine, about 50 feet above chute 10, the grooves on the polished walls plunge 70° S.   In the JoJo, a small prospect on East Willow Creek near North Creede, the vein strikes north of west and dips about 65° S.   Striae on a slicken-sided surface plunge about 85° W.   In the Amethyst vein striae and grooves plunge about 70° S., indicating that the hanging wall moved toward the south as it descended.   At some places striae on the walls plunge steeply north, and in the Amethyst mine striae plunging north were observed within a few feet of those plunging south.

All these observations are in accord with the generally accepted theory that in regions of normal faulting the dominant element of movement is downward.   At only one place in this district have striae as flat as 45° been observed.   The hanging wall of the Amethyst fault has moved with a zigzag motion, at one time north, then south, but the vertical element of movement, so far as indicated by striæ, was greater than the horizontal element.

CHARACTER OF THE ORES.

COMPOSITION AND METALS CONTAINED.

The shipping ores of the Amethyst lode are highly siliceous. In general they carry from 60 to 85 per cent of silica, but some shipments run as low as 45 per cent. Barite is nearly always present in the shipping or partly oxidized ore. In the main these ores carry from 10 to 20 per cent of barium sulphate; one lot of 1,026 tons from the Bachelor mine, shipped in August, 1906, carried 22.1 per cent. Aluminum, as a rule, ranges from 3 to 8 per cent, but in some shipments it will reach 13 per cent. Lime is surprisingly low. In analyses of over 100 lots it averages less than 1 per cent, and only one analysis shows as much as 2 per cent. In most of the ore that is shipped sulphur runs less than 2 per cent, although in some shipments it is as high as 4.5 per cent. In the concentrating ores obtained from deeper levels it is higher. Many shipments of rich ore carry less than 5 per cent of iron and manganese, and they rarely run over 10 per cent, although some shipments carry as much as 13 per cent. In the ores that are shipped zinc in general amounts to less than 2 per cent.

The silver contents in large lots ranged from 20 to 70 ounces to the ton in ores shipped between 1902 and 1912. A few shipments contained as little as 14 ounces. Lead ranges from 1 per cent or less to 5 per cent or a little more. Some of the richest silver ores carry less than 1 per cent of lead, and some ores with 6 to 7 per cent of lead carry less than 20 ounces of silver to the ton. A close scrutiny of over 100 analyses does not show that silver increases or decreases with any other metal. There is also no clear relation between silica and silver. Some of the most highly siliceous ores, carrying over 80 per cent of SiO_2, yield less than 25 ounces of silver to the ton. Although the baritic ores carry in the main more silver and less gold than the chloritic ores, the proportion of barite in the ores may vary greatly without a corresponding variation in silver. In the partly oxidized ores the variation in silver appears to be independent of the variation in other minerals.

On the south end of the Amethyst lode, in the Bachelor and Commodore mines, gold is almost negligible. In the shipping ores from the New York, Last Chance, and Amethyst mines the gold content ranges from 0.03 to 0.25 ounce to the ton, the average being near 0.1 ounce ($2). In the concentrating ores from the Happy Thought mine the gold runs 0.1 to 1.0 ounce to the ton. Some of the lead concentrates from the Happy Thought mine carry from 1 to 3.7 ounces of gold.

The concentrating ores from the north end of the Amethyst lode, from the lower levels of the Amethyst mine, and from the Happy

Thought mine contain only a few ounces of silver to the ton. The concentrates obtained carry sphalerite, anglesite, cerusite, pyrite, and gold, with a little chalcopyrite. The gangue is quartz and chlorite with very little barite. No analyses of these ores are available.

The ores of the Solomon-Holy Moses lode in the lower levels are similar to the chloritic ores of the Amethyst lode but carry more lead and zinc and less gold. Silver is only sparingly present. Some rich silver ore was taken in the early nineties from the Holy Moses workings near the surface. It was this ore that first attracted attention to the Creede district, but the total production of silver from the Solomon-Holy Moses lode is small compared with that from the Amethyst lode. Barite was present in the Holy Moses silver ore but was not as abundant as in the silver ores of the Amethyst lode.

(The ores of the Alpha-Corsair lode are similar to those of the upper levels of the Amethyst lode. They carry little or no gold, however, and thuringite, if present, is not abundant. Only a little lead is present and only 1 to 2.8 per cent of zinc. Silver ranges from 20 to 65 ounces to the ton in analyses of 45 lots shipped since October, 1902. Silica ranges from 58 to 80 per cent, iron and manganese from 2.9 to 6.5 per cent, aluminum oxide from 6.9 to 16 per cent. Lime averages about 1.0 per cent. No analysis shows more than 2.8 per cent of zinc in ore shipped prior to 1912. Barite is reported in only three shipments and is less abundant than in the Amethyst vein. Sulphur averages about 3 per cent.

The average value of the Creede ores is probably about $28 a ton. Some of the ore mined in the early nineties carried more silver. An estimate made in 1892 places the value of the ore mined in that year at $100 a ton. The total production of the Commodore mine, which included some material of relatively low grade, averaged 44 ounces of silver to the ton. Much of the ore from the Last Chance and New York mines was milled from the Big stope. The deposit was found in rock cut by closely spaced parallel fractures, and it is certain that much country rock was included in the ore. Nevertheless the ore was of high grade and was probably somewhat richer than that of the Commodore mine.

Some rich ore of the Amethyst vein carried as much as 1,000 ounces of silver to the ton, but no records of large shipments of such ore are available. A review of numerous assay sheets and all the other data that are at hand indicate that 50 ounces to the ton appears to be near an average figure for the richer ores of the upper part of the lode.

In the lowest levels of the mines and on the Wooster adit the silver decreases notably. Some ore runs as high as 20 ounces a ton or higher, but the ratio of gold and zinc to silver is much higher than the ratios in the higher levels. Not much of the ore in the lower levels carries more than 12 ounces of silver to the ton, although the gold, zinc, and lead bring the ore in some of the stopes on the lower levels well within the workable limit. Some of this ore has been enriched slightly by secondary processes.

STRUCTURE AND PARAGENESIS OF THE PRIMARY ORES.

Some of the ores have been deposited in open spaces and some by replacement of the wall rock. It is not everywhere possible to distinguish ore that has been formed by one process from that formed by the other, and where replacement has been almost complete there is but little evidence remaining of the older material that has been replaced. It is believed, however, that the major part of the deposits were formed in open spaces. The chloritic ore in the lower levels is thought to have filled openings in the main, but some has replaced the wall rock. In this ore the sulphides are arranged here and there in sheets, but at many places they occur as small bodies scattered irregularly through the chloritic gangue. Some of the ore exhibits crustified banding. A piece of such ore from the Amethyst lode is shown in figure 6. The specimen is about 5 inches long and 4 inches wide. On the outer margin on either side is a layer of thuringite, quartz, sphalerite, and galena, an association which is typical of the

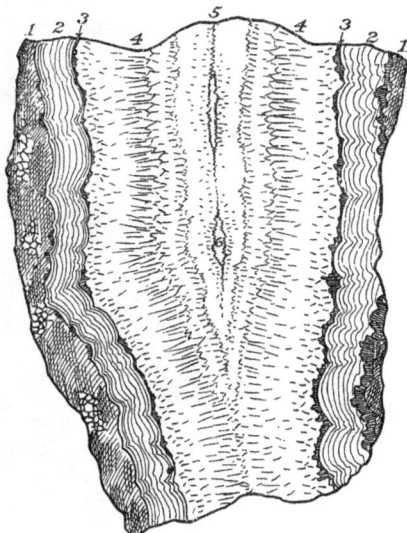

FIGURE 6.—Crustified ore from Amethyst lode. 1, Thuringite, quartz, sphalerite, and galena; 2, finely banded quartz; 3, sphalerite and a little quartz; 4, amethystine quartz; 5, union of quartz combs; 6, druse.

lower levels in both the Amethyst and Solomon-Holy Moses mines. The second layer from the outside is one of finely banded quartz. This is about half an inch thick or a little thicker and is approximately uniform in thickness and appearance. Next to it, toward the middle, is a thin layer of sphalerite with a little quartz. On one side this layer is not everywhere developed. Within this layer toward the middle are two layers of amethystine quartz 1 inch thick or a little thicker. These unite at one end of the specimen, but at the other end there is between them a layer of white comb quartz. At one end of this layer is a small vug representing an unfilled space in the vein. This specimen is noteworthy in that the same sequence of layers is shown from both sides toward the middle. This is not an isolated example, for the same sequence was noted in several specimens from the Amethyst dump.   It is possible that the amethystine quartz and associated minerals have been deposited as a veinlet cutting the chloritic ore. This inference is suggested by the fact that the amethystine quartz is practically barren of the precious metals, whereas the chlorite ore generally carries gold and silver.   Amethystine quartz is found, however, interbedded with chlorite (thuringite) in some of the ore, and these veinlets also may have been deposited on chlorite without intervening fracturing.

Brecciated zones are common along the lodes.   An example is shown in figure 7, which is a cross section of the Amethyst vein on level 11 in the Last Chance mine, 175 feet south of the shaft.   This ore carries about 18 ounces of silver and 0.1 ounce of gold to the ton

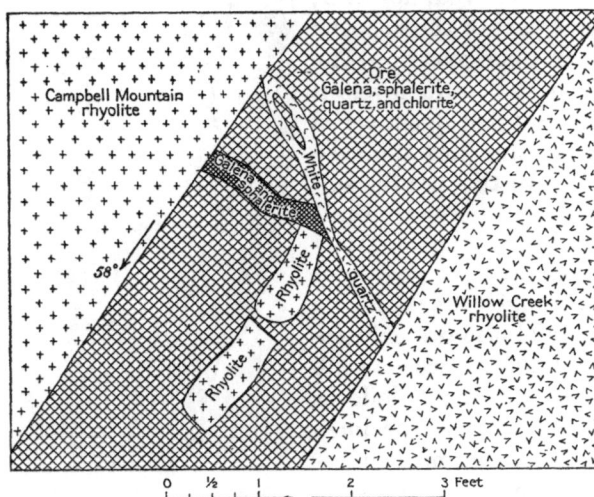

FIGURE 7.—Section of Amethyst vein on level 11, Last Chance mine, 175 feet south of shaft, looking north.

and 8 per cent of lead.   The earlier ore contains angular fragments of rhyolite and is cut by later veinlets of galena and sphalerite and of white quartz.   No barite is present.   At this place, and also in the Overholt mine, as shown by figure 8, the fragments of brecciated rhyolite are angular.   The solutions that cemented these materials apparently had little power to dissolve the rhyolite, for the edges of the blocks are sharp.   At some places there has been movement and brecciation during the deposition of the ore.   Thus, in the Amethyst vein, on level 6 of the Amethyst mine, 300 feet from the shaft, the broken fragments of quartz in sulphides and some of rhyolite are surrounded by gouge and crushed quartz and subordinate sulphides, as shown by figure 9.   Some of the sulphides show indistinct banding, and some of the sulphide ore is brecciated and cemented by quartz (fig. 10).

A banded texture of sulphides with thuringite and quartz is shown in the stopes of the Amethyst mine above level 9, 120 feet north of the Amethyst shaft. Next to the footwall the ore consists of quartz, galena, and sphalerite, stained slightly with limonite. This layer is

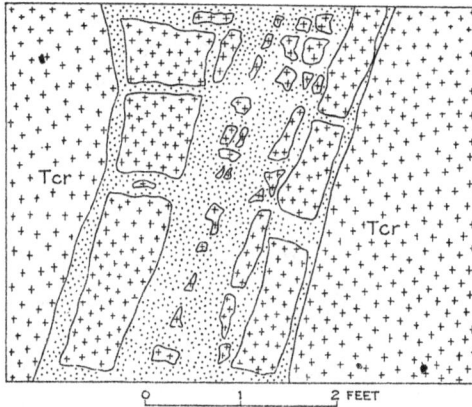

FIGURE 8.—Section of Overholt vein 190 feet above Nelson adit. Tcr, rhyolite; stippled areas, barite and sulphides.

from 3 to 6 inches thick. Above it is one of quartz and limonite about 4 inches thick, and above this in turn is 2 feet of thuringite, quartz, galena, and some sphalerite. At some places the oxidized ore and sulphide ore are found together in small stopes, oxidation having taken place apparently along the fracture or group of fractures. In

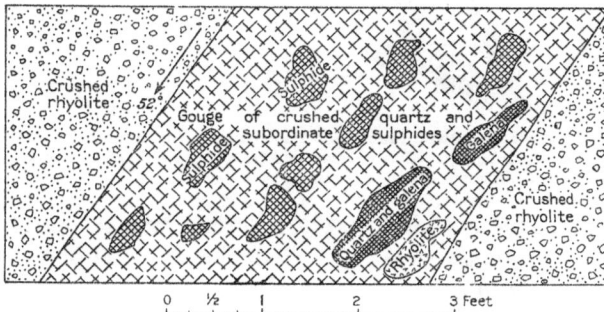

FIGURE 9.—Section of Amethyst vein, Amethyst mine, level 6, 300 feet south of shaft, looking north.

the Solomon mine, in the stopes about 40 feet above the Solomon adit, near chute 16, 3,000 feet from the portal of the Solomon adit, the thoroughly oxidized ore is found as a tabular mass lying between bodies of the sulphide ore, which is almost unaltered. This point is more than 1,000 feet below the surface, but nevertheless some of the ore is almost completely oxidized.

The paragenesis of some of the native silver ore is noteworthy.   A polished specimen of such ore from the Commodore mine is shown in figure 11.   This specimen was not found in the course of this investigation, but it is stated on good authority that it represents a fairly common type in the richer oxidized ores mined in earlier years. The dark lines represent native silver; the areas above and below are chalcedony, red jasper, and white quartz.   The quartz, which alternates with silver and jasper, is clearly crustified, and it appears certain that all these minerals were deposited in open spaces.

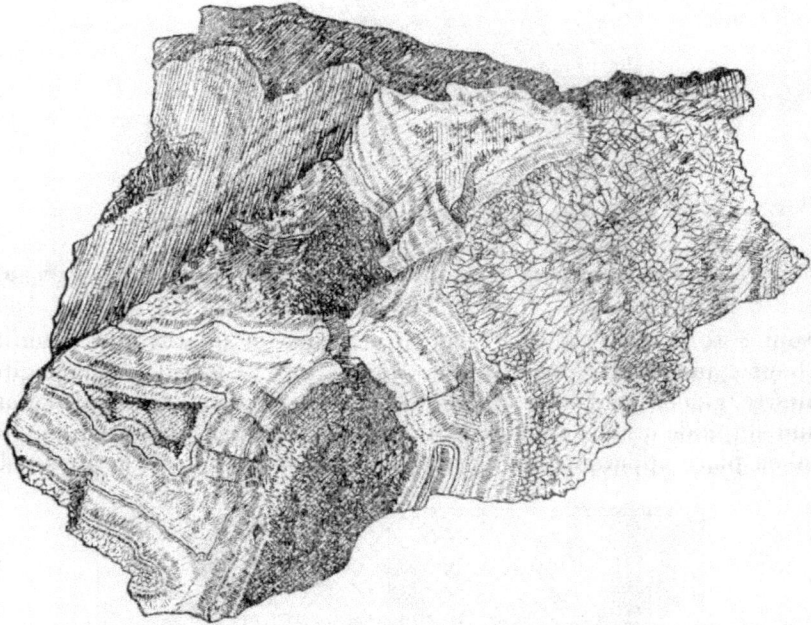

FIGURE 10.—Brecciated ore cemented by quartz, Amethyst mine. The dark areas are fragments consist . ing of sphalerite, galena, and quartz. These are surrounded by banded milky quartz on which amethys-tine quartz is deposited. A vug is shown in the upper right-hand corner.

Native silver is almost invariably regarded as a secondary mineral in deposits of sulphide ores.   This conclusion is supported by the occurrences of the native metal in only the upper portions of nearly all silver veins.

The ores of the Monte Carlo and Mollie S. mines differ from those of the Amethyst and Solomon-Holy Moses veins.   There is no evidence of crustification, but nearly all the ore is shattered rhyolite cemented with carbonates and silicates of copper that carry some silver and gold.   Some of the silver is present as argentite and cerargyrite.   Carbonates are not abundant, but one specimen from the Monte Carlo mine shows small crystals of siderite crusted on the surface of rhyolite and coated over with small crystals of clear quartz.

The genesis of these ores is uncertain; no corresponding zones of sulphide ores have been developed below them. The Mollie S. and Monte Carlo ores have been deposited in shattered zones, whereas the Amethyst and Solomon ores have been deposited in large fault fissures.

The paragenesis of the oxidized ores and the deposition of secondary gold ores with manganese ore are discussed on pages 130–132.

HYDROTHERMAL METAMORPHISM.

The Willow Creek rhyolite nearly everywhere forms the footwall of the Amethyst lode, and it forms both walls of the Solomon-Holy Moses lode on the lower levels. This rhyolite, which is described on page 19, is a purple, drab, gray, or lilac-colored flow, nearly everywhere fluidal, and contains small light-colored bands of porous material at many places. It is a normal rhyolite in composition. Its aphanitic groundmass contains numerous crystals of orthoclase and biotite and a few of plagioclase. A little white mica and a carbonate are present in some specimens. Apatite was not noted. Small dots of iron oxide are found here and there, and the color of the rock is doubtless due to the presence of iron in the groundmass.

FIGURE 11.—Polished surface of ore, Commodore mine. The dark band is native silver; it is surrounded by quartz and jasper.

The Campbell Mountain rhyolite, which overlies the Willow Creek rhyolite, is a reddish-brown or drab flow breccia spotted with lighter-colored inclusions. Although in the field it shows fairly constant differences from the Willow Creek rhyolite, due to its different color and texture, it has approximately the same mineral and chemical composition. It is described on page 25.

Along the lodes these rhyolites have been altered by the vein-forming solutions. Where they are fractured small veinlets of quartz have been deposited, and other veinlets are found composed of pyrite, galena, and sphalerite. Quartz also appears in veinlets of sulphides near the veins. The groundmass of the rhyolite is altered by hydrothermal processes to quartz and chlorite. Sphalerite, galena, and pyrite are deposited in veinlets in the shattered rock and are sparingly disseminated through it. A little sericite is present but only a little. Compared with wall rocks of many other

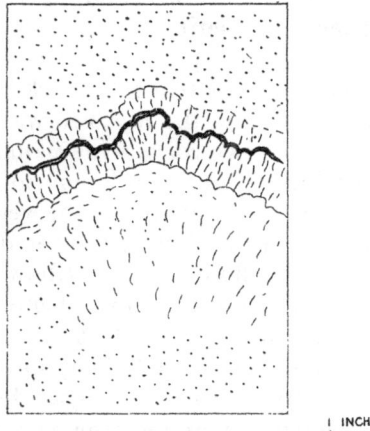

metal-bearing districts of the West, those of the Creede veins show surprisingly little sericitization. At some places the wall rock within 2 feet of the vein is essentially unaltered.

The intrusive porphyry that occurs at several places on the Amethyst lode has not been found in a perfectly fresh condition. In its altered phases hornblende crystals have altered to chlorite, magnetite, and other minerals, and chlorite has formed extensively in the groundmass. Pyrite is disseminated through the rock, and dots of magnetite are abundant in the groundmass as well as in the hornblende crystals.

In both the Amethyst and Solomon-Holy Moses lodes much of the filling consists of a green gougelike chloritic material, laminated in places by pressure and movement. It contains locally abundant galena, sphalerite, and pyrite. This green material also replaces the wall rock, but practically everywhere it is disturbed by movement, so that it is impossible to mark off exactly where the filled part of the vein ends and the replaced wall rock begins.

Of the two rocks whose analyses are given below specimen 1 was obtained from the Solomon mine, on the adit level between chutes 9 and 10. At this point the vein is about 600 feet below the surface and the rock is but little altered by oxidation. It is a purple rhyolite, a typical flow of the Willow Creek rhyolite. The groundmass is purplish and contains phenocrysts of feldspar and a little mica. Along this vein 700 feet farther north, 50 feet above the adit, at a depth of about 1,000 feet, specimen 2 was obtained. This specimen is the typical green gouge which makes up the greater part of the Solomon vein and a large part of the Amethyst vein. It is composed of chlorite (thuringite), quartz, and some white chalky material, shown under the microscope to be altered rhyolite. The rhyolite carries high silica and potash; but its soda is very low (0.63 per cent). The green gouge carries much less silica but approximately the same alumina. In the formation of the gouge there was a notable decrease in potash and a decrease also in soda but a large increase in ferrous iron, magnesium, and water and a noteworthy increase in manganese. Powder of this rock, which is too soft to section, is composed principally of chlorite with some quartz and chalky decomposed rhyolite, and here and there a flake of sericite. Microscopic study of the powder of specimen 2 shows no adequate source of magnesia except the chlorite.

Briefly the hydrothermal processes at Creede are characterized by the development of much thuringite and a little sericite.

Analyses of rocks from Solomon mine.

	1	2		1	2
SiO_2	73.53	55.25	TiO_2	0.19	0.15
Al_2O_3	12.87	12.10	CO_2	.23	.11
Fe_2O_3	.88	1.28	P_2O_5	Trace.	Trace.
FeO	.64	10.71	S	.02	.11
MgO	.56	9.30	Cr_2O_3	None.	....
CaO	.07	.34	MnO	.09	1.43
Na_2O	.63	.28	BaO	.05	Trace.
K_2O	8.92	.39			
H_2O-	.40	1.49		99.78	99.64
H_2O+	.70	6.70			

1. Willow Creek rhyolite, lower adit, between chutes 9 and 10.
2. Green thuringite gouge, lower adit, 50 feet above chute 16.

CHARACTER OF SOLUTIONS THAT DEPOSITED THE PRIMARY ORES.

The composition of the solutions that deposited the primary ores is not known. Something may be inferred of their character, however, from the materials which they deposited. The solutions probably carried alkaline chlorides, alkaline carbonates, alkaline sulphides, and sulphates. At Wagonwheel Gap,[47] 12 miles southeast of Creede, hot springs are now issuing, probably from a fissure that carries barite and fluorite, with small amounts of gold, silver, and copper. Analyses of these waters show that the solutions are mixtures of chlorides, carbonates, and sulphates. They carry silica, alkalies, lime, magnesia, iron, aluminum, and a little lithium. The vein shows clear evidence of crustification and, like the deposits of Creede, carries much barite. Fluorite, which is abundant at Wagonwheel Gap, is only sparingly present at Creede, but a little has been noted in material from the upper levels of the Solomon-Holy Moses mine.

It is believed that the solutions that deposited the ores of Creede were mixtures of alkaline chlorides, carbonates, sulphides, and sulphates, and that they carried abundant silica, with lead, zinc, iron, and a little copper, together with gold and silver. Barium was present, probably as sulphate. They probably carried alumina also, for thuringite occurs in the ore and replaces the wall rock along the lode. Magnesia and manganese were deposited in appreciable quantities in the chloritic ore. The rhyolites that form the walls of the veins carry very little lime and magnesia. The ores are practically free from lime, but the chloritic ores carry considerable magnesia. Either the waters carried originally large proportions of magnesia or they dissolved magnesia from rocks lower than the rhyolites that now inclose the deposits.

---

[47] Emmons, W. H., and Larsen, E. S., The hot springs and mineral deposits of Wagonwheel Gap, Colo.: Econ. Geology, vol. 8, p. 235, 1913.

AGE AND GENESIS OF THE DEPOSITS.

The faulting of the region involves the Creede formation and is therefore, at least in part, later than that formation, which is of Miocene age, probably upper Miocene. As the ores have been deposited in and along the fault fissures, the ores themselves are Miocene or later.

The latest rocks that occur in the immediate vicinity of the more valuable deposits are the quartz latite porphyry dikes. These rocks have been altered at some places by the vein-forming solutions and are therefore older than the ores that were deposited by those solutions. Moreover, the porphyry has been faulted and is therefore older than some of the faults. However, in point of age the porphyry is more closely correlated with the deposition of the ores than other rocks of this region.

The ores have been deposited in open spaces in the fissures and have replaced the wall rocks along the fissures. As the deposition of the ores was associated in time and place with volcanism, and as the mineral deposits are like those that are known to have been deposited elsewhere by ascending hot waters, it is believed that the ores of Creede have been deposited by ascending hot waters. The source of the solutions is not known, but as the period of deposition of the ores may be most closely correlated with that of the intrusion of the porphyry, it is probable that the ores are genetically related to the deeper rock masses that supplied the material for its intrusion. Such a relation is suggested, though it has not been proved. It is noteworthy that faulting is more extensive in the Creede district than in the surrounding region. It is believed that the faults extend to considerable depths and that they have facilitated the ascension of waters from the deeper sources.

SUPERFICIAL ALTERATION OF ORES.

The unaltered ore of the Creede district is composed mainly of quartz, barite, thuringite, sphalerite, galena, and pyrite. Rhodochrosite and chalcopyrite are very sparingly present in some deposits. Native gold is also primary. Silver is associated with galena and doubtless also with other sulphides. Some of it occurs as argentite.

When attacked by oxygenated waters, the sulphide ores are altered to minerals that are comparatively stable in an oxidizing environment. These include limonite, hematite, jarosite, wad, chalcedony, kaolin, gypsum, anglesite, smithsonite, malachite, and other minerals.

The outcrops of the ore deposits are stained yellow or brown with limonite. A little hematite is present also, and jarosite was noted at the outcrop of the Amethyst vein between the Amethyst and Last Chance shafts. The outcrops are not highly ferruginous, however, although there are considerable bodies of ferruginous ore at lower levels—for example, on level 6 of the Amethyst vein north of the Amethyst shaft. Limonite forms from the oxidation of pyrite and thuringite, and very subordinately from the oxidation of chalcopyrite. Neither pyrite nor chalcopyrite is abundant, although pyrite is generally present in the sulphide ore. Chalcopyrite is comparatively rare. The shipments of zinc concentrates carry considerably more zinc than iron, and on the concentrating tables much of the pyrite is saved with zinc. Some of the limonite has doubtless formed from the oxidation of thuringite. This mineral, as shown by the analyses on page 101, carries 40.93 per cent of iron oxides. In the unoxidized ore of the Solomon-Holy Moses lode and on the north end of the Amethyst lode thuringite is much more abundant than pyrite.

Manganese dioxide (wad) is present in most of the oxidized ores. It is a decomposition product of thuringite, which according to the analysis on page 101 contains 2.75 per cent of MnO, and very subordinately of rhodochrosite. In the upper levels wad is much less abundant than limonite, but in some of the deepest levels, where thin fractures are coated or filled with oxides, wad is more abundant than limonite.

Quartz is comparatively stable in the oxidized zone. Some silica is dissolved, however, as is shown by analyses of waters from the Amethyst and Solomon-Holy Moses mines. Moreover, silica has been deposited by the underground waters as chalcedony and quartz, which are locally crustified with native silver. Barite is abundant at the outcrops of the Amethyst vein and is present also in outcrops of most of the other productive lodes of this region, as well as in a number of unproductive lodes. Some of the barite is secondary. The rhyolites contain considerable barium. On weathering this is dissolved as carbonate, and in the presence of sulphate waters secondary barium sulphate has doubtless been deposited. Barite is only sparingly present in the deepest levels of the Solomon-Holy Moses lode and in the deeper levels of the mines at the north end of the Amethyst lode. It is possible that the primary ore carried originally more barite in upper parts of the lodes than in the deeper levels.

Galena oxidizes to anglesite and cerusite. Anglesite, the most abundant secondary lead mineral in the Creede deposits, occurs both massive and as crystals (fig. 12). Much of the high-grade oxidized lead-silver ore is a mixture of anglesite and cerusite carrying nodules and tabular bodies of unoxidized galena. Claylike gouge, presumably altered rhyolitic material, taken 310 feet south of the Last Chance shaft on level 7, carries 5.42 per cent of PbO and 3.23 per cent of SO_3. As was pointed out long ago by Penrose,[48] the first

[48] Penrose, R. A. F., jr., Superficial alteration of ore deposits: Jour. Geology, vol. 2, p. 297, 1894.

stage of the alteration of galena in many deposits produces anglesite, which subsequently changes to cerusite. This conclusion is drawn from the fact that nodules of lead sulphide are surrounded by shells of lead sulphate, which in turn are surrounded by shells of lead carbonate.[49]

In the lower part of the oxidized zone in the Amethyst mine, anglesite is much more abundant than cerusite and apparently has altered directly from galena. Along thin fractures in galena anglesite has developed by processes of oxidation, as is shown in figure 13. At higher levels nodules of galena are covered with shells of anglesite, which in turn are covered with cerusite and kaolin. One of these nodules (fig. 14) from the dump of the Last Chance mine was studied in some detail. The inner sphere of galena, about 2 inches in diameter, is covered with a shell of dense solid anglesite 0.3 inch thick. The galena shows the characteristic cubical cleavage. There is a distinct banding of the anglesite, which is intricately crenulated, especially near the outer margin. The sulphate is so firmly attached to the sulphide that a thin sliver may be broken across the contact without separating the two minerals. A little carbonate is present

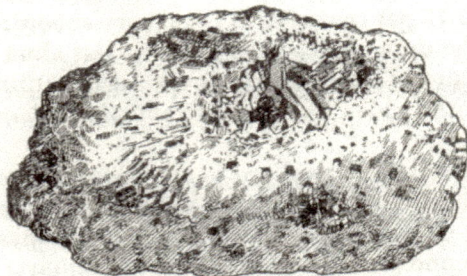

FIGURE 12.—Massive anglesite with vugs containing anglesite crystals, Amethyst mine.

near the outer surface of the nodule, where it is associated with some kaolin.

A chip of anglesite to which a small mass of galena adhered was broken from a nodule and boiled in colored water. Near the outer margin of the specimen, where anglesite is probably altering, some thin porous layers became apparent after washing. Another porous layer was disclosed exactly at the contact between galena and anglesite. Tests show that this layer is anglesite, not cerusite. There is every reason to suppose that this porous layer is migrating inward as the sulphide nodule slowly changes to sulphate, and that the dense glassy anglesite of the outer margin of the specimen has been formed from the spongy anglesite shell, which has doubtless moved from the outside of the specimen 0.3 inch or more toward the center. The outer dense shell appears to represent, therefore, the cementation product of earlier spongy shells.

According to present theories of solution, during its oxidation the lead sulphide is at first dissolved. Only a small amount of lead

---

[49] Emmons, S. F., Geology and mining industry of Leadville: U. S. Geol. Survey Mon. 12, p. 546, 1880.

sulphide can be dissolved in cold water. According to Weigel 3.6×10^{-6} mols, or 0.000861 gram, dissolves in 1 liter at 18° C. Oxygenated waters convert the sulphide in solution to sulphate: $PbS + O_4 = PbSO_4$. The solubility of lead sulphate is likewise low; 0.041 gram is dissolved in a liter at 18° C.[50] By comparing the porous material next to the galena with the dense material of the margin, it is obvious that the oxidation of galena does not take place molecule by molecule, but solution, transportation, and precipitation are intermediate processes. Some of the lead sulphate is precipitated probably in place, forming the spongy shell, and some lead sulphate moves outward to cement the outer part of the sponge.

Sphalerite alters to smithsonite and to goslarite. Calamine and willemite have not been identified by the writers in Creede ores. In the ores now remaining in the mines smithsonite is not abundant. Goslarite occurs at many places but is comparatively soluble and does not accumulate as large bodies of secondary ore. Zinc was not determined in the waters that drain from the Amethyst and Solomon-Holy Moses lodes, and, as stated by Chase Palmer,

FIGURE 13.—Galena (dark) altering to anglesite (light) along cracks. Ore from Park Regent mine, Amethyst lode, 1,200 feet below surface.

there was no evidence of its presence. In view of the abundance of sphalerite and the rarity of secondary zinc minerals, it is difficult to account for the disposition of the zinc that must be dissolved in the oxidizing zone in considerable quantities.

Feldspars and rhyolite undergoing oxidation yield kaolin and aluminum sulphate. The sulphate incrusts the walls of mine workings but, being soluble, does not accumulate. The altered rhyolite, of which an analysis is stated on page 125, probably carries a little aluminum sulphate. There is no evidence of the presence of basic ferric sulphate in the sample analyzed. Only a little iron is in the ferrous state, and there is not enough lead oxide to balance the SO_3 for anglesite. The downward-moving waters deposit chalcanthite and melanterite here and there, but neither of these minerals is stable. Gypsum is comparatively rare and has been noted only in oxidized ores.

---

[50] See Smith, Alexander, Introduction to inorganic chemistry, rev. ed., p. 403, 1910.

Copper is almost negligible in Creede ores. Of the total production of 1909, amounting to $1,061,220, only $2,262 was copper. In the records of production of earlier years, when the richer secondary ores were exploited, no mention is made of copper. In the deeper unoxidized ores a little chalcopyrite was noted at several places. It is probably the only primary copper sulphide in the district. In the oxidized ores malachite, chrysocolla, chrysoprase, and chalcanthite occur in very small quantities as oxidation products. No chalcocite was recognized, although it was looked for diligently. Possibly it is present in the sooty black powder that occurs sparingly on ore surfaces in the lower levels.

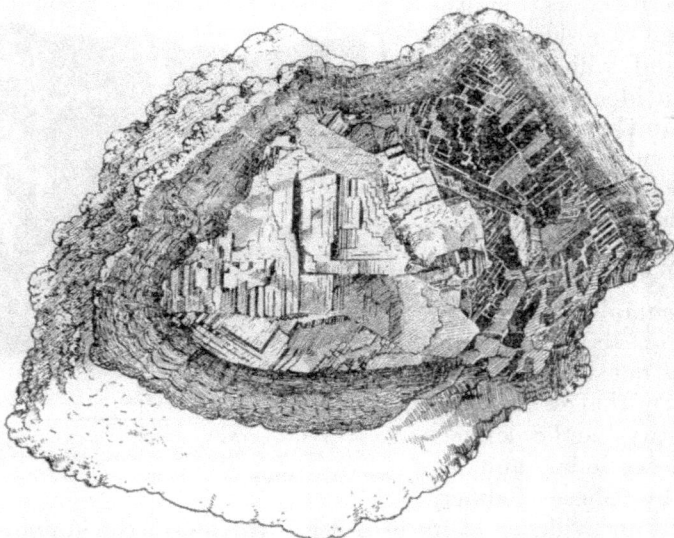

FIGURE 14.—Nodule of galena altering to anglesite, Last Chance mine, Amethyst lode.

Some of the thuringite ores from the lower levels of the Amethyst vein, when treated with hydrochloric acid, give at once a pronounced odor of hydrogen sulphide. So far as the writers know only three mineral sulphides, alabandite, pyrrhotite, and sphalerite, are capable of rapidly generating hydrogen sulphide with acid. Neither alabandite nor pyrrhotite could be detected in the ore, and the crystals of sphalerite that were isolated did not give enough hydrogen sulphide to be detected by its odor. Moreover, after the chlorite ore is treated with acid a second time the odor is only faintly perceptible, and after a third treatment no odor was noted. The source of the hydrogen sulphide could not be determined. Possibly it is present as minute bodies of alabandite in galena.[51]

---

[51] Nishihara, G. S., The rate of reduction of acidity of descending waters by certain ore and gangue minerals and its bearing on secondary sulphide enrichment: Econ. Geology, vol. 9, pp. 746–747, 1914.

The table below gives analyses of a sample taken below the oxidized zone in the Amethyst vein (No. 1) and another taken on level 7 of the Last Chance mine, 310 feet south of the shaft (No. 2). Sample 2 is altered oxidized gouge near the middle of the vein. The rhyolites are not uniform in composition, nor is all the altered ore like the gouge analyzed. It can not be safely assumed that sample 2 was derived from material like sample 1. A comparison of the two is useful, nevertheless, for it is believed to show the general nature of the superficial alteration. This, however, varied somewhat from place to place. The altered sample No. 2 probably contained originally more of the metals than sample 1. Alumina and lime are approximately the same in both samples. The altered rock has less silica. It contains more iron, sulphur trioxide, phosphorus pentoxide, and lead. These changes may be expressed in terms of minerals. Sample 1, calculated as the norm, shows a little corundum and hypersthene, the elements of which in the rock are in micas. The norm composition calculated for sample 2 is probably near its actual composition as regards the principal minerals. By the oxidizing processes large amounts of orthoclase and plagioclase are decomposed. Some of the silica is released as quartz; much kaolin is developed from feldspar; iron sulphides are changed to melanterite and copiapite; pyromorphite and anglesite are formed from galena. Water is added to form hydrous minerals.

Analyses of rhyolite and of rhyolite gouge in oxidized zone, Amethyst vein.

	1	2		1	2
SiO_2	77.36	67.06	TiO_2	0.16	0.09
Al_2O_3	11.37	11.69	CO_2	.06	None.
Fe_2O_3	.31	2.11	P_2O_5	.03	.71
FeO	.36	.25	SO_3		3.23
MgO	.14	Trace?	S	.33	
CaO	.30	.33	MnO	.03	Trace.
Na_2O	1.38	.45	BaO	.05	
K_2O	7.28	.81	PbO		5.42
H_2O-	.55	2.37			
H_2O+	.26	5.52		99.97	100.04

1. Willow Creek rhyolite, Nelson adit.
2. Rhyolite gouge in oxidized zone, level 7, Last Chance mine.

Norm of rhyolite and mineral composition of clay gouge, Amethyst vein.

	1	2		1	2
Quartz	39.9	49.5	Apatite		0.6
Orthoclase	43.9	5.0	Limonite		.8
Albite	12.1	3.7	Melanterite		.9
Anorthite	1.1		Copiapite		4.8
Kaolin		25.5	Pyromorphite		2.7
Rutile	a .3	.1	Anglesite		4.2
Corundum	.5		Water	0.8	2.3
Calcite	.1		Other radicles	.3	
Hypersthene b	.3				
Pyrite	.6			99.9	100.1

a Calculated as ilmenite.   b Not in either rock.

DOWNWARD ENRICHMENT.

GENERAL CHARACTER.

In many deposits of copper, silver, and gold the ore near the surface is of relatively low grade and unworkable. At greater depths richer ore is encountered, and at still greater depths rich ore gives way to lower-grade sulphides. The leached portions of the deposits are in general highly oxidized, whereas the lower-grade ore in the deeper portions of the deposits shows comparatively little alteration. These relations are generally assumed to indicate that the metals have been dissolved from the upper portions of the ore bodies and have been redeposited lower down.

To state it briefly, many deposits may be divided into four zones— (1) an oxidized leached zone at or near the surface, (2) an oxidized or partly oxidized zone of richer ore below the leached zone, (3) a zone of rich sulphides below the rich oxidized ore, and (4) a zone of lower-grade primary sulphide ore below the rich sulphides.[52]

The theory of downward enrichment assumes that underground waters in advance of erosion have dissolved the valuable metals from the outcrops and carried them to lower depths, where they are precipitated in an environment that is not so highly oxygenated and therefore less acid. By these processes the valuable metals from the highly oxidized and superficial zones and from the portions of the lodes that have been worn away by erosion are reconcentrated at greater depths, where they may form bonanzas. In some districts, however, erosion has been more rapid than thorough leaching, and rich ores of gold and silver may occur at the surface, or gold placers may be formed from the waste of lode material that is eroded. Rich ore was found at the outcrop of the Solomon-Holy Moses vein on the Holy Moses claim and in the Amethyst vein on the Last Chance and Amethyst claims.

On the Amethyst vein (see fig. 15) the richest ore near the outcrop was at a relatively low place in the lode, presumably at points where erosion has been more rapid than on either side. Placers from the Creede ore deposits have not been recognized, although the deposits carry appreciable quantities of gold. The ores richest in gold are not at the surface but below it.

Figure 15 is a vertical projection of the Amethyst vein with curvatures eliminated. Workings are projected on the plane from the vein that dips south of west about 55°.

The richest ore had been mined before this investigation was undertaken, but from studies of ore remaining, of stope sheets of the mines, and of authentic data supplied by several officials of the mining

---

[52] Emmons, S. F., The secondary enrichment of ore deposits: Am. Inst. Min. Eng. Trans., vol. 30, p. 177, 1900.

companies the position of the richest ore may be approximated with a fair degree of accuracy. In general, the zone from the surface to a depth of 150 to 300 feet is not productive, although the surface ores on the Amethyst and Last Chance claims, where the lode crops out at lower levels, were richer than those on the Commodore and Happy Thought claims. Stopes were raised to the surface on this part of the lode. Near the north end line of the Commodore claim a pit marks the apex of a cave that extends downward several hundred feet. Some very good ore and much barren rock was taken from this cave by milling from below, but figures showing the grade of the ore near the surface of the cave are not available. In general, the stopes that are raised in the upper zone shown on figure 15 are run only a few feet because the mining was found to be unprofitable.

The "most productive zone" shown on figure 15 is the bonanza zone of the lode. It supplied most of the ore mined in the 10 years after the discovery (over $20,000,000 worth) and also much of the

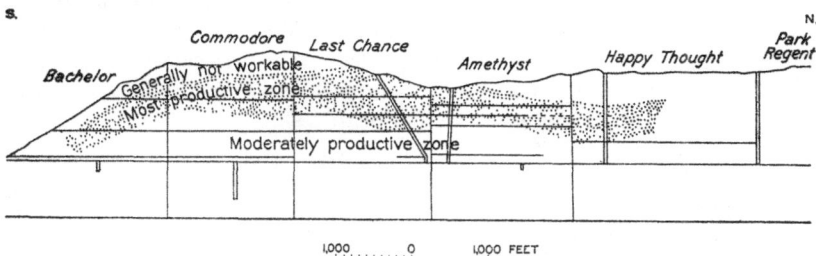

FIGURE 15.—Vertical projection of the Amethyst lode with curvatures eliminated.

ore that was mined in the following decade. A large proportion of the shipping ore produced to-day also comes from this zone. Not all of the lode included within this zone is of workable grade, but perhaps more than half of it is, and if all the workable ore had been removed it would probably be possible to pass through stopes from the Bachelor mine to the Happy Thought mine. In general, the ore of this zone is oxidized, although partly oxidized sulphides are abundant, especially in the lower levels. Native silver, cerargyrite, and gold are the principal minerals. The moderately productive zone carries ore that consists mainly of lead and zinc sulphides in a siliceous, chloritic gangue. The silver and gold contents are much less than in the zone above, but some of the ore of the lower zone is of shipping grade and much of it is concentrating ore. Along thin fractures anglesite, limonite, and manganese dioxide have formed in the ore of this zone also, but comparatively little of the ore is oxidized.

There is no evidence of much sulphide enrichment in ores of the Solomon-Holy Moses lode, which carry lead and zinc as their principal metals, with subordinate amounts of gold and silver. According to reports, however, silver ore of good grade was taken from upper levels

of the Holy Moses mine. It is said that the richest ores of the
Corsair-Alpha mine were found above or not far below the water
level, within a few hundred feet of the surface.

FRACTURED CONDITION OF THE LODES.

Most of the lodes of the district have been fractured since the
primary ores were deposited. The evidence of late fracturing is particularly clear on the Amethyst lode. Strong fracture planes, generally slickensided, follow one or both walls at many places.
Locally—for example, in the Amethyst mine—the sulphide ore itself
is polished smooth as glass. At one place the polished sulphides show
two sets of striae, indicating movement in at least two directions
after the ores were deposited.

There are few bodies of ore in the Amethyst fault that have not
been shattered or at least fractured. At some places the ore is little
more than an uncemented or imperfectly cemented breccia, and some
ore has been mined by drawing it out of lower openings without
blasting. On the lower levels the ore is commonly fractured, but
less intensely than in the higher levels. The conditions for rapid
downward circulation of water are particularly favorable. The lower
adits driven on the Amethyst, Solomon-Holy Moses, and Alpha-
Corsair lodes carry streams of considerable size.

MINE WATERS.

The agent of enrichment of the ores is the surface water that soaks
into the lodes, dissolves the metals, and carries them downward to
be deposited at lower levels where air is excluded and the valuable
metals are precipitated. Owing to the openings incident to mining,
the water circulation is doubtless more vigorous to-day than it was
before the lodes were discovered, and doubtless the waters are more
highly oxygenated also, especially in the upper levels. The analyses of
mine waters indicate, therefore, only the general character of the
solutions that are assumed to have altered and enriched the ores.

The analyses of two samples of water from the mines are stated
below. Both samples are from streams of considerable size, and each
represents the waters that are drained from one of the veins.
Each sample was placed in two 1-gallon bottles, and only a small
air space (about 5 cubic centimeters in each gallon bottle) was left
unfilled in order as far as possible to prevent oxidation.

Analyses of mine waters from Creede, Colo.

[Parts per million. Chase Palmer, analyst.]

	1	2		1	2
Na	28.52	7.5	SO_4	60.12	104.4
K	2.41	3.5	Cl	Trace.	1.6
Ca	17.36	46.2	HCO_3	62.83	73.5
Mg	1.30	7.3	Fe_2O_3 (Al_2O_3)	2.10	1.2
Mn	.25	3.2	SiO_2	32.35	23.2

1. Amethyst vein, Bachelor mine, lowest adit, 1,400 feet from portal.
2. Solomon vein, Solomon mine, lowest adit, 1,560 feet from portal.

Concerning these analyses Mr. Palmer makes the following statement:

In each sample a slight iron deposit was filtered off before analysis. Neither copper nor hydrogen sulphide was found. The samples showed evidence of neither lead nor silver. Zinc was not looked for specially, but there was no evidence of its presence. The dissolved iron was in the ferrous state. Both waters are alkaline. To ascertain definitely the presence of zinc, silver, and lead, larger samples would be necessary.

Both of these waters are from depths of about 1,400 feet below the surface. After warming in the sun they were clouded slightly with light-gray precipitates, and notwithstanding precautions against oxidation both developed thin deposits of iron oxide before they were analyzed at Washington. Both are alkaline and both carry iron in the ferrous state. This, however, was determined as ferric oxide, as stated in the analyses. Chlorine is present in both waters but abundant in neither. Both carry a little manganese. If any aluminum is present, it is included with iron. Silica is present in appreciable amounts in both waters. They carry sulphates of alkalies and alkaline earth with acid carbonate and considerable excess carbon dioxide not reported above.

COMPOSITION OF ALTERED ORES.

In the lower levels silver is present in galena, sphalerite, and pyrite and in quartz that is stained dark gray by reason of thinly disseminated sulphides. Whether the silver molecule is intergrown with the lead sulphide as an isomorphous mixture or whether it is in argentite has not been determined. In the oxidized zone silver occurs mainly as silver chloride and as native metal. Argentite has been noted in the Mollie S. ore and is doubtless present also as a secondary mineral in the ore of the Amethyst lode. Stephanite is reported to be present in the Alpha-Corsair lode but was not found during this investigation. Some complex sulphide ore with a mere trace of antimony was found, but the mineral could not be identified. None of the complex antimony or arsenic sulphosalts of silver were found in the ore of the Amethyst lode. In view of the fact that these minerals are present in a majority of the large deposits of silver ore in the middle and late Tertiary rocks of the United States, their absence in the Amethyst lode is noteworthy. It may be that they have been overlooked, but rather careful scrutiny of cabinets containing numerous specimens of the richer ores taken out during the days of mining the bonanzas failed to reveal any of these minerals. Two hundred analyses from smelters do not record the presence of antimony or arsenic.

Pyrargyrite and proustite, as shown by Irving and Bancroft,[53] are abundant in the ore deposits of Lake City, Colo. According to

[53] Irving, J. D., and Bancroft, Howland, Geology and ore deposits near Lake City, Colo.: U. S. Geol. Survey Bull. 478, p. 62, 1911.

Ransome [54] proustite is present in many deposits in the Silverton quadrangle. Neither of these minerals has been noted by the writers in any of the ores of Creede, although they have been looked for with some care. In the lower portion of the zone of enriched ore argentite and possibly some other silver salts may have been present in considerable quantities, but in the main the silver in enriched ore occurs as native metal and as chlorides, rather than as sulphides.

SECONDARY PRECIPITATION OF SILVER AND GOLD.

Ground water containing sulphuric acid dissolves silver, and solution is increased by the presence of ferric sulphate.[55] As the solutions descend, silver as native metal is precipitated. This may be brought about by the accumulation of ferrous sulphate and by decrease in acidity. Cooke has shown that silver sulphate may be held in solutions containing varying proportions of ferric sulphate and ferrous sulphate, but when concentration of ferrous sulphate proceeds beyond a certain rather small amount, silver will be precipitated as native metal.[56]

Some of the silver of the Amethyst vein has been precipitated directly rather than by interaction with solids. Beautiful specimens of silver are intergrown with chalcedony and quartz. In some of this ore the chalcedony contains enough iron oxide as thinly disseminated particles to be colored red. Much of the richest ore mined from the Amethyst vein is said to have been red jasper containing abundant native silver. Apparently iron in some oxidized form has been precipitated along with the native metal.

Fairly concentrated silver sulphate solutions reacting with carbonates deposit native metal after acidity has been reduced. In the Amethyst vein carbonates are not abundant. A little rhodochrosite is present, but it is doubtful whether these reactions have been of great importance there.

H. C. Cooke,[57] Palmer and Bastin,[58] and F. F. Grout[59] have shown that silver sulphate solutions reacting with several sulphides will precipitate silver. Palmer and Bastin's experiments were carried out with solutions of silver sulphate, and those of Grout and Cooke with silver sulphate carrying considerable excess acid. Sullivan[60]

---

[54] Ransome, F. L., A report on the economic geology of the Silverton quadrangle, Colo.: U. S. Geol. Survey Bull. 182, p. 82, 1901.

[55] Cooke, H. C., The secondary enrichment of silver ores: Jour. Geology, vol. 21, p. 10, 1913.

[56] Idem, p. 18.

[57] Idem, p. 21.

[58] Palmer, Chase, and Bastin, E. S., Metallic minerals as precipitants of silver and gold: Econ. Geology, vol. 8, p. 140, 1913.

[59] Grout, F. F., On the behavior of cold acid sulphate solutions of copper, silver, and gold with alkaline extracts of metallic sulphate: Econ. Geology, vol. 8, p. 417, 1913.

[60] Sullivan, E. C., The interaction between minerals and water solutions, with special reference to geologic phenomena: U. S. Geol. Survey Bull. 312, pp. 37, 64, 1907.

has shown that silver will be precipitated by many of the silicates. It is thrown down from the sulphate solution by orthoclase, albite, amphibole, clay gouge, and many other substances. Chalcocite and other minerals containing cuprous copper act very readily with sulphuric acid solutions to precipitate silver.[61] Thuringite precipitates metallic silver from silver sulphate.

Precipitation of silver in the Amethyst vein has probably been brought about in various ways—by precipitation in open cavities owing to the accumulation of ferrous sulphate, or by reactions with thuringite, feldspar gouge, and sulphides.

Sulphuric acid solutions carrying chlorides in the presence of manganese dioxide dissolve gold very readily. Downward-moving waters transport the gold to lower levels and in depth it is precipitated by ferrous sulphate, bicarbonates, sulphides, and many other materials.[62]

So many substances precipitate gold from chloride solutions that its descent will be retarded unless some oxidizing agent is present to delay or inhibit precipitation. This is accomplished by manganese oxides, which will change ferrous sulphate to ferric salt. In the Happy Thought and Amethyst mines the evidence is very clear that gold has been precipitated from downward-moving waters. The richer ores of the mines are found in thin fractures with manganese dioxide. Gold and manganese have been precipitated together, doubtless owing to decrease in the acidity of the solutions.[63] All the rich gold ore in the Creede district is associated with manganese oxide, which has presumably been reconcentrated by downward-moving waters from material in which gold occurs sparingly. The unoxidized ore carries in general about $1 or $2 in gold to the ton; the richer ore may carry $10, $20, or locally $100 to the ton or even more. If an attempt is made to separate the gold from manganese dioxide it accumulates in the bottom of the pan, with considerable lead sulphate and galena. In a spoon of horn the gold may be overlooked, but by using an ordinary gold pan the gold may easily be separated from the lead minerals. Particles as large as mustard seed are frequently found. They are dark yellow and apparently of a high degree of fineness.

As shown by the study of the annual production and of numerous smelter returns, there is a noteworthy increase in gold with depth. The higher proportion of silver near the surface and the increase in

---

[61] Palmer, Chase, and Bastin, E. S., op. cit., p. 155. Grout, F. F., op. cit., p. 416.

[62] Emmons, W. H., The agency of manganese in the superficial alteration and secondary enrichment of gold deposits in the United States (paper read at Canal Zone meeting, Am. Inst. Min. Eng., 1910): Am. Inst. Min. Eng. Trans., vol. 42, p. 4, 1911; The enrichment of sulphide ores: U. S. Geol. Survey Bull. 529, p. 125, 1913; The enrichment of ore deposits: U. S. Geol. Survey Bull. 625, p. 306, 1916. Brokaw, A. D., Secondary precipitation of gold in ore bodies: Jour. Geology, vol. 21, p. 251, 1913.

[63] Brokaw, A. D., op. cit., p. 258.

the proportion of gold with depth are due in part, however, to the accumulation of the silver chloride and native silver in the upper levels. Evidence is clear in some places that gold has been carried downward to greater depths than silver. Thus in the Amethyst mine, on level 11, a stope of sulphide ore carrying manganese and gold in thin fissures assays $15 a ton, whereas the ore in general in the lower part of the mine carries only $1 or $2 a ton in gold and only a few ounces of silver.

EXTENT OF AMETHYST VEIN BEFORE EROSION.

Many veins carry bonanzas near the surface. In some veins of late age the bonanzas are large and rich, and there is doubt as to their secondary origin, because the amount of vein matter that has been eroded from above their present outcrops appears to be insufficient to have supplied the metals that the bonanzas contain, if it is assumed that the original vein material was of a grade similar to that now found in depths below the bonanzas. But this is not true of the Amethyst vein. The throw of the Amethyst fault is at least 1,400 feet, and nearly everywhere along this fault there is little difference between the elevation of the hanging wall and that of the footwall. At least 1,400 feet of the footwall of the Amethyst fault has been eroded after the fault was formed. The veins were probably formed soon after the faulting took place, and if so at least 1,400 feet and probably much more of rock has been eroded from this region since the Amethyst vein was formed. The material that has been removed by erosion seems to be sufficient to have supplied all the rich ores that have been mined from its upper portions, if it is assumed that the rich ore is all of secondary origin and derived from the concentration of metals from a lode of even width carrying ore that was originally no richer than that now found in the lower levels of the Amethyst vein.

If it is assumed that all the valuable metals that have been leached from eroded portions of a lode were precipitated in the same lode at lower levels, and that all the metals leached from the outcrop and the low-grade zone below it have likewise been precipitated in depth, a simple formula [64] may be applied to obtain an estimate of the amount of erosion since the ores were deposited. Although this formula does not take into account the changes in mass in the ore itself, the pore space developed, nor the scattering of metals that are dissolved, these factors are probably not important enough to modify the result very greatly.

The vertical extent in feet (x) of the part of the lode which has been removed from above the present apex can be computed by the following equation, in which a equals the vertical extent in feet of

---

[64] Emmons, W. H., The enrichment of sulphide ores: U. S. Geol. Survey Bull. 529, p. 47, 1913.

the leached zone, b the vertical extent in feet of the enriched zone, l the assay contents (stated in any convenient unit, as dollars or ounces per ton) of the material remaining in the leached zone, e the similar contents of the enriched ore, and p the similar contents of the primary ore:

$$x = \frac{a\ (l-p) + b\ (e-p)}{p}$$

Lodes that have been deposited by ascending hot waters in middle and late Tertiary time, like those of Creede, have formed relatively near the surface. As the precious metals are rarely deposited by hot waters in workable quantities at or very near the surface, it appears probable that at Creede a certain proportion of the lode material that has been eroded was of low grade or barren. A correction for this condition would indicate that the results obtained by working out the equation given above are too small. On the other hand, many deposits formed relatively near the surface contain rich ores that are doubtless in part of primary origin, deposited not at the surface but near it. Any correction made for this condition would decrease the estimate. If it is assumed that corrections applied for these two conditions are of the same magnitude and thus offset each other, the solution of the equation will give a fairly accurate estimate of the amount of material eroded from above the vein.

The leached zone above the "most productive zone" in the south end of the Amethyst lode is about 200 feet deep on a vein dipping 50° to 60°. This would give from 225 to 240 feet as measured on the incline. But near the Amethyst shaft there was no "leached zone," for rich ore was found at the surface. Let the leached zone, measured on the incline, be assumed to have a downward extent of 225 feet. The grade of the ore in this zone is not known, yet no one familiar with the mines and the ore market in this region would place the tenor much above 10 ounces of silver to the ton. This ore carried very little gold. The most productive zone, as shown by figure 15, has a range in its vertical extent from 400 to 575 feet at most places, although locally its range is a little more. On the incline this is about 600 feet. This zone, however, contained certain lower-grade portions, and if allowance is made for these portions, 400 feet of the vein measured on the incline appears to be a fair estimate for the enriched zone. The ore mined from this zone was rich in silver, and for reasons stated on page 112, 50 ounces to the ton appears to be a reasonable estimate for its silver content. For the solution of the equation the following figures may thus be assumed: $a=225$ feet; $b=400$ feet; $l=10$ (ounces of silver to the ton); $e=50$; $p=12$.

$$x = \frac{225(-2) + 400(38)}{12} = \frac{-450 + 15,200}{12} = 1,229$$

A vein 1,229 feet above the present apex measured on a dip of 55° would be about 1,025 feet vertically above the present apex. As stated on page 88, at least 1,400 feet of rock has been removed by erosion since the Amethyst vein was formed. It is clear that there has been sufficient erosion in this region for the vein matter from above the Amethyst vein to supply all the metals now concentrated in that vein, even if the ore removed by erosion from above the present apex was somewhat poorer than the material in lower levels driven in the vein or if the part of the vein eroded was a little narrower than the average width of the part of the vein now developed.

GASES IN AMETHYST LODE.

When the Humphreys tunnel was extended northward from the bottom of the Park Regent shaft subterranean gases were encountered in considerable quantity and it was necessary to employ a large exhaust pump to afford suitable working conditions for the miners. The working face of this tunnel is more than 2 miles from the portal but only about 1,000 feet from the bottom of the Park Regent shaft, through which ventilation is provided. The part of the tunnel between the shaft and the face is driven on the vein, which is a wide brecciated zone filled with green chlorite gouge and sulphides. The lode is there about 1,400 feet below the surface. Oxidization of chlorite to iron oxide and of galena to anglesite has taken place along small fractures, but less than 1 per cent of the ore is oxidized.

This part of the mine was first visited by the writer on a clear day, and he suffered no serious inconvenience through the presence of noxious gases, although it was necessary to burn several candles at once in order to get adequate light. Later, on a stormy day, an attempt was made to revisit the face, and at a point only 400 feet in from the Park Regent shaft he found difficulty in breathing, his head began to ache, and he felt a sensation of dizziness. It was necessary to return hastily to the Park Regent shaft to prevent collapse. No work in extending the tunnel in this part of the mine had been done for nearly two years, and it was surprising that the gases should still be so abundant.

A study of these gases was made in 1903 by Harry A. Lee.[65] At that time the Humphreys tunnel had been driven some hundreds of feet north of the Park Regent shaft but not so far as it is now. Lee notes that the gas is lighter than air and is found in the top parts of the workings. It is a colorless, warm gas and by some is said to have a sweetish taste. A sample collected by Lee and J. W. Finch was analyzed by W. F. Edwards:

---

[65] Lee, H. A., Gases in metalliferous mines: Colorado Sci. Soc. Proc., vol. 7, pp. 163–188, 1903.

Analysis of gas from north end of Humphreys tunnel on Amethyst vein.

[W. F. Edwards, analyst.]

	Per cent of volume.
Nitrogen by difference	96. 08
Oxygen by pyro absorption	3. 92
Carbon dioxide	0
Carbon monoxide	0
Nitric oxide	0
Oxygen	0
Hydrogen	0
Ethylene	0
Methane	0

Nothing absorbed by KOH, CuCl, $KMnO_4$, Br in H_2O, and no explosive or combustible substance was found.

It is said that the flame of a candle placed in front of a crevice in the rock will be drawn toward the crevice in clear weather and that it will be drawn away from the crevice in cloudy weather. High barometric pressure seems to cause a flow from the workings into the fractures in the rock and low pressure a flow from the rocks into the mine workings. For this reason the end of the adit is nearly clear of gases except on stormy days, when the barometer or air pressure is lower.

The physiologic effect of these gases is presumably simply that of suffocation. It is safe to enter the workings as long as candles will burn in them; but caution should be used as to entering when a large kerosene torch will burn and candles will not. An acetylene lamp will burn with even less oxygen than a torch, and therefore its use as an indication of safe conditions is attended with some risk.

At Cripple Creek likewise the gases in some of the mines have a suffocating effect and deaths have resulted in workings imperfectly ventilated. These gases are discussed by Lindgren and Ransome.[66] Below are given some analyses for comparison.

Analyses of subterranean gases and of air, by volume.

	1	2	3
Carbon dioxide	9. 25		
Nitrogen a	82. 80	96. 08	79. 06
Oxygen	7. 95	3. 92	20. 94

a By difference.

1. Average of two samples collected by L. C. Graton from Conundrum mine, Cripple Creek, Colo. A. W. Browne, analyst.
2. Sample collected by H. A. Lee and J. W. Finch from Humphreys tunnel, Creede, Colo. W. F. Edwards, analyst.
3. Amounts by volume of principal constituents of air, after Ramsay. Clarke, F. W., The data of geochemistry, 4th ed.: U. S. Geol. Survey Bull. 695, p. 42, 1920. 0.93 per cent of argon is included with the nitrogen. Air commonly contains also water vapor and about 0.03 per cent of carbon dioxide.

---

[66] Lindgren, Waldemar, and Ransome, F. L., Geology and gold deposits of the Cripple Creek district, Colo.: U. S. Geol. Survey Prof. Paper 54, p. 252, 1906.

Three hypotheses may be suggested to account for the gases in the Amethyst lode—(1) they may be the residual air from which oxygen was abstracted by oxidation of mine timbers and other organic material; (2) they may be residual air from which oxygen was abstracted by oxidation of sulphides and gangue minerals; (3) they may have been confined in fractures in the rocks before the mines were opened.

The first hypothesis may be set aside because the part of the lode from which the gases come has little or no timbering, and timbered portions at higher levels are free from similar gases. Moreover, the oxidation of timbers would yield carbon dioxide, and although that compound might be removed from the residual gases by the carbonation of minerals, it is unlikely that all of it would be so removed. Moreover, carbonates are very sparingly present in this part of the lode. The Amethyst ores in depth carry only a small fraction of 1 per cent of carbonates.

The oxidation of minerals would remove oxygen from the air, and in the residuum the proportion of nitrogen would be increased. As noted above, the mine gases have the composition of air from which the large part of the oxygen has been removed. In this portion of the Amethyst lode oxidation is not extensive, but a small fraction of the ore is oxidized. Lead sulphate has formed from galena, and chlorite is altered to iron oxide. It is possible that the gases in the Amethyst lode represent the residuum of entrapped air.

On the other hand, some of the gas may have been confined in fractures in the rocks before the mines were opened. The rocks containing the ores of the Creede district are of Miocene age, and the ores are Miocene or later. Volcanic processes have been active in this district in comparatively late geologic time. Possibly the subterranean gases represent an end phase of volcanism. A similar hypothesis in regard to the origin of subterranean gases at Cripple Creek is entertained by Lindgren and Ransome.[67] There is one essential difference, however, between the gases at Cripple Creek and those at Creede. At Cripple Creek an appreciable amount of carbon dioxide is present, whereas none is recorded in analyses of the gases of the Amethyst lode. Removal of oxygen from air might give a gas like that in the Amethyst lode but not such a gas as that collected at Cripple Creek. Nearly all volcanic gases contain at least a little carbon dioxide, although some contain less than 1 per cent.[68] The evidence is not conclusive and is in part conflicting. It is not certain whether the gases that issue in workings on the Amethyst lode are of volcanic origin or whether they are residual from air that has oxidized ore and gangue minerals.

---

[67] Lindgren, Waldemar, and Ransome, F. L., op. cit., p. 257.
[68] Clarke, F. W., The data of geochemistry, 4th ed.: U. S. Geol. Survey Bull. 695, p. 278, 1920.

EXPLORATION OF DISTRICT.

In considering the future development of a district that contains lode deposits the geologist may consider four groups of possibilities— (1) extensions on the strike of developed deposits, (2) fractures and veins in the walls of developed deposits, (3) extensions in depth of developed deposits, and (4) undiscovered lodes. Although the geologist may disclaim any prescience of undeveloped ore bodies in a district, he may with propriety direct attention to what he regards as the most promising localities to seek for them.

EXTENSIONS ON STRIKE OF DEVELOPED DEPOSITS.

As the Amethyst has been the principal productive lode in the district, possible extensions along its strike are of great interest. The geology of the Amethyst fault system is described on page 88. The lode is now developed underground for about 10,500 feet along its strike on the Nelson adit level. Northward the Nelson-Wooster-Humphreys tunnel is driven about 1,000 feet north of the Park Regent shaft. At the face of this tunnel the fault is strongly mineralized and the altered zone is of good width, although the metal contents are relatively low. No ore is known to be developed on the Amethyst fault system between the face of this tunnel and the Equity mine. That mine exploits a vein which is not on the Amethyst fault but which probably intersects it near the portal of the Equity. This vein occupies a fault fissure which is regarded as belonging to the Amethyst system. The distance between the face of the tunnel on the Park Regent vein and the Equity is about 15,000 feet. For about 6,000 feet of this distance, immediately north of the north end of the tunnel, there are no deep developments, and at most places the country is thickly covered with drift. A few shafts were put down in this stretch in the early development of the country, but these were long ago abandoned, and to judge from the size of their dumps and the character of the material elevated, it is doubtful whether they actually reached the lode. For reasons that are stated on page 91 it is believed that the Amethyst fault passes not far from the Captive Inca shaft, but whether or not it is encountered in the Captive Inca workings is uncertain. In any case there is probably more than a mile of the Amethyst fault between the Park Regent and Captive Inca that is concealed by morainal material on the surface and essentially unexplored underground. Between the Captive Inca and Equity mines, also, the Amethyst fault is essentially unexplored.

The southern part of the Amethyst fault, from the south end of the Commodore vein through the Bachelor, is exposed only here and there. Where exposed it generally carries vein matter, but no ore has been mined from it. The output of the Bachelor mine has been derived mainly from veins approximately parallel to and in the footwall of the Amethyst fault. The point farthest south where the Amethyst fault is exposed underground is in the Nelson adit about 600 feet from its portal. There the material from the fault is crushed and altered to a soft gougelike mass. It is iron stained and said to carry small amounts of metals, but it contains no ore. Southeast of this point, in the region of North Creede, there is a complex of faults. One of these, probably the one with greatest throw, crosses the south end of Campbell Mountain not far north of the junction of the two forks of Willow Creek. It has a throw of about 1,400 feet, which is comparable to that of the Amethyst fault. In the Jo Jo Tunnel the fault contains a little quartz and is said to carry small quantities of precious metals. The fault continues its course southeastward and crosses the south end of Mammoth Mountain. The Pipe Dream Tunnel, on Dry Gulch, is probably on the same fracture or fault zone. Two tunnels are driven on this fault on Mammoth Mountain. Only one of these, the Mammoth tunnel, was accessible when visited. It contains quartz, barite, and oxidized metalliferous minerals. Near the point where the fault crosses the ridge of Mammoth Mountain the dumps of workings now inaccessible reveal altered rock and some vein matter.

Within the last few years very little work has been done in searching for the northwest and southeast extensions of the Amethyst lode. In view of its great productiveness, the belief appears to be warranted that prospecting on the Amethyst fault system offers greater promise of reward than that in some other parts of the Creede region where prospecting has been carried on more vigorously. The extensions of the Solomon-Ridge-Holy Moses and Corsair-Alpha veins are treated on pages 94-96.

FRACTURES AND VEINS IN WALLS OF DEVELOPED DEPOSITS.

Between the Bachelor and Park Regent mines many small mineralized fractures make off from the Amethyst fault. In the Bachelor mine these are principally in the footwall; in the Commodore and Last Chance-Amethyst they are mainly in the hanging wall. Where such fractures are closely spaced large ore bodies may result. Indeed, the richest part of the Amethyst lode was in the New York-Last Chance and Delmonte section, where, owing to the presence of many closely spaced fractures in the hanging wall the width of the lode was expanded to nearly 100 feet. The production and character of the Amethyst lode appear to justify a more thorough prospecting of its walls, especially of the hanging wall.

EXTENSIONS IN DEPTH.

As bearing on the possibility of profitable developments below the Nelson adit level, inquiry may be made respecting the position of this level in the zone of enrichment. If the adit is below the bottom of the altered zone any ore bodies that may be encountered below the adit are primary. If the adit is above the bottom of the zone of alteration, bodies of enriched ore may be expected.

With respect to the distribution of silver it may be stated that no bonanzas comparable to those which were worked in the upper levels of the mines have been developed within 100 feet of the adit level. Stopes have been raised at many places from the adit, but they are small compared with those higher in the vein, and the ore in them is generally of lower grade. The evidence of metallic content alone, however, does not warrant the conclusion that the bottom of the zone of possible enrichment has been reached. The state of the ore with respect to alteration processes is important in this connection. At the southeastern portion of the vein, in the Bachelor and Commodore mines, comparatively thorough oxidation extends to greater depths than in the mines on the lode farther north. At some places a considerable amount of oxidation has taken place in the Bachelor mine within 100 feet of the adit, and on the Commodore claim, which lies just north of the Bachelor, a stope called the Wire Silver stope was raised from level E, which is just above the Commodore tunnel. The bottom of this stope is about 225 feet above the Nelson drainage adit and about 1,200 feet below the surface. Although this stope was not accessible to the writers, the presence of wire silver in the ore is regarded as an indication of secondary alteration. A winze that was sunk below the adit in the Bachelor mine is now filled with water, but according to report some sulphide ore of shipping grade was developed.

The fracturing of the lode is pronounced, and the secondary fracturing is complicated, the ore being much more permeable at some places than at others. On the whole, however, the circulation of water from the surface is exceptionally vigorous, owing to the highly fractured condition of the ore. These observations lead to the conclusion, therefore, that the adit level in the Bachelor and Commodore mines is probably not below the bottom of the zone which is marked by partial alteration of the ore. Whether processes of enrichment have gone far enough to make exploration of this part of the lode profitable is a question that can not now be answered.

The northwest end of the lode is not so thoroughly oxidized in the lower levels as the southeast end. Nevertheless the sulphide ore in the lower levels of the Last Chance, Amethyst, and Happy Thought is crossed by numerous fractures in and along which limonite and manganese oxide have formed. Some of the ore that is cut by stringers of these oxides is low-grade concentrating ore. Locally, however, this ore is of considerably higher grade, the increase being mainly gold. On level 12 in the Amethyst mine the ore 350 to 600 feet north of the shaft is composed of galena, zinc blende, pyrite, chalcopyrite, and other minerals and is said to carry about 11 per cent of lead, 6 to 8 ounces to the ton in silver, and about $1.20 to $3 to the ton in gold. In another stope about 800 feet north of the shaft and just below level 12 similar ore containing conspicuous veinlets of manganese oxide carries from $5 to $15 to the ton in gold. In the Happy Thought mine, 370 feet north of the bottom of the shaft, the sulphide ore only 20 feet above the adit level is cut by veinlets of manganese oxide. This ore is said to carry 8 per cent of lead, about $1.80 to the ton in gold, and a little silver.

There is not much doubt that the gold in this ore has been increased somewhat by secondary processes, for the association of gold with manganese dioxide is significant of such processes.[69] For example, in the Happy Thought mine on the Amethyst vein between levels 7 and 8 a body of partly oxidized ore composed of galena, zinc blende, copper carbonates, cerusite, and anglesite carries a considerable amount of manganese dioxide, which coats the older sulphides and occurs in fractures cutting the partly oxidized ore. A considerable tonnage of this ore concentrated in the Humphreys mill yielded $20 in gold to the ton, and some of it ran as high as $100 to the ton. This figure is of significant magnitude, for the average content of gold in the Happy Thought ore is much less.

In reviewing these facts it is pertinent to inquire whether they warrant the expectation of any enrichment in gold below the adit level in the north end of the Amethyst vein. It has been shown that the gold in these relations is found in stopes just above the adit level, but so far as is now known the enrichment in gold due to secondary deposition at this level is not great. The present methods of concentration, however, permit the recovery of lead, silver, zinc, and gold, and in much of the ore the gold is subordinate in terms of value. Silver is comparatively low in these levels, but so far as is indicated by the data available there is no reason to suppose that lead and zinc will be less abundant in the zone 200 feet below the adit than in the zone 200 feet above it, provided, of course, that the ore of the vein maintains an equal width.

In connection with the possibilities of profitable development below the adit level certain considerations other than the changes in the character of the ore in depth merit attention. The amount of water that issues from the portal of the adit is large, and most of it is collected in that part of the adit which is driven on or near the lode.

---

[69] Emmons, W. H., The agency of manganese in the superficial alteration and secondary enrichment of gold deposits in the United States: Am. Inst. Min. Eng. Trans., vol. 42, p. 3, 1912.

No notable quantity of water is added in the portion of the adit that crosscuts the country rock. In sinking a deep winze that was put down in the footwall in the Commodore mine pumping charges were high, and they would doubtless have been higher if the crosscuts had been run to the vein. Owing to the highly fractured condition of the vein, any project that contemplates deep exploration should provide for handling a large proportion of the water now draining from the adit. Some of this drainage could probably be kept out of lower workings, but extensive stoping below the adit would surely increase the flow.

The possibilities of tunnels below the Nelson adit have, of course, been considered. A crosscut driven north for 5,500 feet from a point near the mouth of Windy Gulch should encounter the lode at a depth between 250 and 300 feet below the Nelson adit. This would not provide for dumping, however, and would necessitate a long extension of the track for disposal of waste. The expense of such an undertaking could be decreased by the development of cheap electric power, but the writers do not profess to know whether the possibilities of finding profitable ore below the adit level would warrant the necessary outlay of capital. A deep shaft sunk below the present adit level would make it possible to explore a larger portion of the Amethyst lode, and if adequate pumping facilities were provided such a project would seem more promising of success.

PROSPECTING FOR NEW LODES.

In many regions where ores have been deposited in igneous rocks the country rock is highly altered along the lodes by the solutions that deposited the ores. In some regions this alteration extends far from the workable deposits. It serves as a guide for exploration, because deposits are more likely to be discovered in an area of altered rock than in one of unaltered rock. In the Creede district alteration is not conspicuous a few feet from the veins, and at some places the country rock within a foot of the veins is essentially unaltered.

In some of the deposits of the Creede district the ore is brecciated and the fragments are cemented by later ore (fig. 10). Nevertheless, the deposits are believed to be of essentially the same geologic age and younger than the faulting of the region. The faulting was probably later than the extrusion of the latest volcanic series. The ore deposits are therefore later than any of the Tertiary rocks of the region. None of the hard-rock formations can be classed as barren simply on account of its age, as all were present when the deposits were formed. Although the principal ore deposits have been found thus far in rocks of the Potosi volcanic series, it is possible that deposits may be found in the later Tertiary rocks. Thus faults in any of these rocks may be considered worthy of investigation as possible seats of ore deposition.

CHAPTER XI.—MINES.

BACHELOR MINE.

The Bachelor mine is about 1¼ miles north of Creede, on the south end of the Amethyst vein. The deposit was first discovered about July 1, 1884, by J. C. MacKenzie, H. M. Bennett, and James A. Wilson. It is stated that some low-grade ore was taken out from the shallow workings and carried by pack animals to Sunnyside, where it was treated without success in an arrastre. Not much work was done until 1891, when, with the revival of mining that followed the discoveries of that year, prospecting was vigorously resumed. The total production is estimated at about $2,000,000, practically all in silver.

The mine is exploited from four adits that have a vertical range of about 1,175 feet. The lowest of these is the Nelson or Wooster adit, which encounters the vein on the Bachelor claim about 1,931 feet from the portal. Tunnel 4, which is one of the principal adits of the Commodore mine, is about 445 feet above the Nelson adit, and tunnel 3 (Commodore) is 401 feet above tunnel 4. Tunnel 1, the highest tunnel, is 329.6 feet above tunnel 3. The highest point of the vein is about 1,375 feet above the Nelson adit. The Bachelor shaft is sunk 125 feet below the adit, and thus the total development below tunnel 1 is 1,500 feet.

Nearly all the ore of the Amethyst vein is carried through the Bachelor claim. Owing to its favorable geographic position, much of the development work in the Bachelor has been done from adits driven by other companies. The larger portion of the ore from the Bachelor claim was taken out by lessees.

The vein crops out at the portal of tunnel 1, near and at the portal of tunnel 3, and a few feet west of the portal of tunnel 4. At these outcrops the vein material is nowhere of paying grade but consists of pink and white barite with a little iron oxide and some quartz. The outcrop near the portal of tunnel 4 is especially conspicuous. Here the lode is a sheeted zone including three or four well-defined fissures, one of which is a fault between the Willow Creek rhyolite and the Campbell Mountain rhyolite. It strikes northwest and dips steeply southwest. The outcrops at the portals of tunnels 1 and 3 are not on the Amethyst fault, both walls at these places being in the Willow Creek rhyolite.

The vein trends northwest from the portal of tunnel 4 to the portal of tunnel 3, where it bends and strikes almost north. Approximately

142

at this portal it is intersected by the Copper vein, which on the surface strikes N. 67° W. and dips 75° SW. The exact point of intersection is not exposed, however. The vein at this elevation is composed of barite, quartz, limonite, hematite, black manganese oxide, some copper carbonate, silver chloride, and native silver. Silver is not so abundant as in lower levels, and barite is more abundant. This part of the vein is reported to contain from 18 to 20 ounces of silver to the ton.

Tunnel 1, from its portal to the end line of the Bachelor, is driven on the vein (fig. 18, p.146). At the portal both walls of the vein are in the Willow Creek rhyolite and the vein dips 72° or more to the west. About 250 feet from the portal, for a short distance along the strike, the vein dips steeply east, but farther north it has its usual dip of about 75° W. This is noteworthy, for no local reversal of dip has been noted on the main fault fissure. Two small intersecting veins join the main vein in the hanging wall. One of these is about 60 feet from the portal, the other about 400 feet. To a point about 825 feet in from the portal both walls are the Willow Creek rhyolite. At this point the vein is cut at a small angle by a fault that strikes a few degrees west of north and dips steeply west. The footwall of the fault is the Willow Creek rhyolite and the hanging wall is the Campbell Mountain rhyolite. North of this point the vein follows this fault. On this level the Bachelor vein shows conspicuous sheeting and contains much barite, with iron oxide, but practically no ore has been mined.

Tunnel 2 is a short adit whose portal is inaccessible. On tunnel 3 the vein is followed for about 1,150 feet to the Commodore mine. For this entire distance both walls are in the Willow Creek rhyolite. Stopes are raised here and there at this level, and considerable ore has been mined between levels 1 and 3. At a point about 100 feet south of the north end line of the Bachelor a level is turned from a raise inclined 70° W. about 78 feet above tunnel 3. About 50 feet north of the top of the raise a 100-foot crosscut is run in the hanging wall. The country rock is the Willow Creek rhyolite. At 80 feet from the main vein is a parallel vein that strikes N. 15° W. and dips into the main vein at a high angle.

In tunnel 4 the vein crops out 20 or 30 feet southwest of the portal. A thin zone of sheeting in the Willow Creek rhyolite is exposed also on the northeast side of the portal. Near the portal the vein, which dips about 58° SW., is not workable. The adit is driven on the fault fissure for about 325 feet, then westward in the hanging wall 276 feet, then northward about 150 feet to the fault fissure. Here the fault strikes N. 58° W. and dips 75° SW. The hanging wall is the Campbell Mountain rhyolite and the footwall the Willow Creek rhyolite. The tunnel, which crosses the fault, encounters the main lode 100 feet farther north, where it is barren, and follows the lode northward beyond the Bachelor into the Commodore mine. Along the strike of the fault, about 450 feet N. 57° W. of the point where it is crossed by the tunnel, it is encountered near the end of a 230-foot crosscut driven west from the tunnel. ˙ At this point the fault strikes N. 54° W. and dips 72° SW. It is mineralized, and a small stope 3 feet wide has been opened. The main lode, which is followed by the tunnel, is about 180 feet from the fault on the footwall side, and between the lode and the fault there are six small stringers. Beyond the crosscut the tunnel follows closely the main lode N. 20° W. to the end line of the Bachelor. Throughout this distance the hanging wall and footwall are the Willow Creek rhyolite. It is supposed that the fault turns beyond its second exposure on level 4 of the Bachelor, its strike changing from N. 54° W. to N. 10° W. The strike of the fault is known to change on the surface and in the Copper lode workings. Moreover, in the Commodore mine, near the south boundary of the Commodore claim, the fault and the vein in the footwall of the fault join again, intersecting at a small angle. South of this point stopes are carried on both the

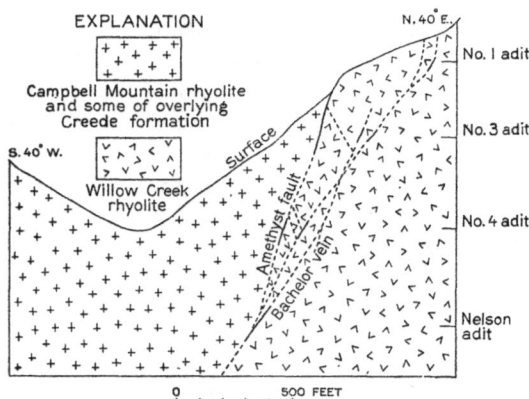

FIGURE 16.—Vertical section in Bachelor mine S. 40° W. from a point 90 feet north of portal of adit 1. Shows Bachelor vein in footwall of Amethyst fault.

fault and the vein, and at the intersection the stopes are much wider than the average. To the north the vein follows the fault fissure beyond the Commodore end line. On this level there is more than 1,000 feet of the fault fissure that has not been explored, and in view of the fact that the fault fissure carries ore at many places and that it could easily be reached by short crosscuts, the chances of finding ore would seem to justify the small outlay necessary for exploration.

At the point where the west crosscut is run to the fault, there is a horse of silicified rhyolite breccia about 4 feet wide and 20 feet long. Stopes are carried above and below on both sides of this horse. The vein below dips 71° W. Some 50 feet above the level the veins join and the stopes are 12 feet wide. Cross sections are shown by figures 16 and 17. Both of these veins are in the Willow Creek rhyolite in the footwall of the Amethyst fault.

Level 5 of the Commodore mine intersects the vein on the Bachelor claim about 400 feet south of the north end line of the Bachelor, and stopes are raised here and there from this level. Both walls of the vein are the Willow Creek rhyolite.

The Nelson adit is the lowest tunnel that encounters the vein. It is driven N. 68° W. as a straight crosscut 2,117 feet long. About 600 feet from the portal it encounters a dike of decomposed porphyry (see fig. 18), through which it is driven for 150 feet. This dike follows the Amethyst fault. The rock east of it is the Willow Creek rhyolite and that west of it is the Campbell Mountain rhyolite. The dike rock is highly crushed and altered to a soft gougelike mass, and it contains some minor iron-stained fractures but no workable ore. About 1,400 feet from the portal a crosscut from the tunnel is turned N. 18° W., and at 525 feet from this turn the tunnel crosses the Amethyst fault and a few feet beyond encounters the lode. The relations, which are somewhat complex at this point, are illustrated by figure 18. Near the junction of the fault with the lode the lode splits and includes a long, thin wedge of rhyolite. As the fault is mineralized the faulting is, at least in

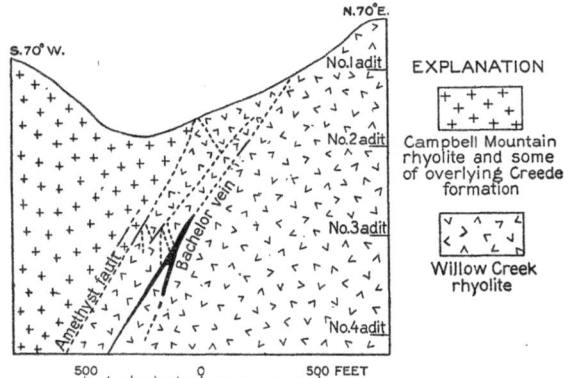

FIGURE 17.—Vertical section in Bachelor mine S. 70° W. from a point 25 feet south of the portal of adit No. 1.

part, older than the period of mineralization, but as the ore is crushed some movement has taken place since mineralization occurred. At the southeast end of the workings on this level the fault is not heavily mineralized but low-grade ore is reported to be present. Small stopes on the fault and on one of the veins in the footwall have been raised in tortuous workings above this level. About 1,000 feet north of this point the tunnel again crosses the fault. Near the Commodore end line the vein and fault join. There the fault dips 74° W. and strikes nearly north. The footwall is the Willow Creek rhyolite, and the hanging wall is the Campbell Mountain rhyolite. At neither of the two points where it is cut by the tunnel does the fault look promising, and this accounts for the fact that more than 1,000 feet of the fault is undeveloped on this level.

In a raise about 435 feet north of the Bachelor shaft and 86 feet above the Nelson tunnel the vein dips 50°–60° W. On an intermediate level about 12 feet north of the top of the raise a narrow vein joins the Bachelor vein from the footwall side.   This vein is exposed also on a higher level about 156 feet above the Nelson adit.   It strikes N. 55° W., making an angle of 30° or more with the main vein. It dips 68° SW. and is followed for 60 feet from the point of inter-section with the main vein, which it does not cross.   It carries in places 1 foot or more of good silver ore, and some small stopes are raised near the junction with the main vein.   The ore is partly oxidized but contains some nodules of lead sulphide.   On this level, about 156 feet on an incline above the adit, a drift is run on the main vein 300 feet south of the raise.   The ore, which is partly oxidized, is poor.   Striae on the vein walls plunge about 75° S.

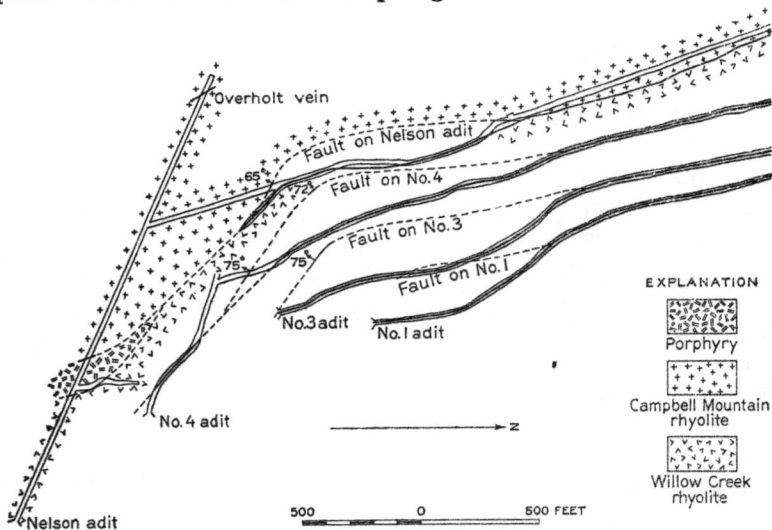

FIGURE 18.—Composite level map of Bachelor mine and part of Commodore mine, showing position of Amethyst fault on four adits.   The drifts for more than 1,000 feet follow a vein in the footwall of the Amethyst fault.

The ore of the Bachelor mine is valuable principally for silver. Gold is present in subordinate quantities compared with that in the ore of the mines at the north end of the vein.   The minerals include quartz, barite, sphalerite, galena, pyrite, chalcopyrite, native silver, and probably argentite.   Oxidation products are limonite, wad, smithsonite, cerusite, and anglesite.   Barite is abundant, especially in the outcrop and at the higher levels.   Analyses of the shipments show 10 to 25 per cent of barium sulphate.   The ore is siliceous, carrying in general from 50 to 65 per cent of silica.   Lead ranges from 1 to 3 per cent; sulphur and zinc from 1 to 4 per cent.   The ore is partly oxidized as deep as exploration goes, and much of it is stained black with manganese oxides.   These are supposed to be derived from thuringite, which is known to carry manganese.   A little rhodochrosite, however, was noted in a vug in ore composed of galena and anglesite. The specimen was taken from the vein in the Nelson tunnel near the point of intersection of the Bachelor vein and the Amethyst fault. Some typical analyses of shipments furnished by the American Smelting & Refining Co. are stated below.

Analyses of ores from Bachelor mine.

[Per cent except as otherwise indicated.]

Date.	Tons.	Ag (ounces to the ton).	Pb.	SiO₂.	Fe, Mn.	Al₂O₃.	Zn.	BaSO₄.	S.
November, 1902.............	366	40.1	2.8	65.6	11	2.8	0.5	13.5	3.9
October, 1903..............	661	22.3	1.2	59.2	2.3	11.9	.3	....	....
November, 1904............	123	45.4	......	59	2.3	11.7	.4	....	.05
June, 1906.................	1,014	20.5	1.3	54.2	9.4	5.6	......	17.1	2

Although the Bachelor mine has produced considerable marketable ore, the ore bodies have not proved so rich nor so persistent as those in the mines to the north. Little or no ore has been stoped above level 1, nor more than 150 feet above level 3. No ore bodies of great value have been found within 200 feet of the surface. As shown by the sections (figs. 16 and 17) the fracturing along the lode is complex. Several of the fractures are locally mineralized, but the principal ore shoots are on the footwall side of the Amethyst fault, in the fissure that is nearly parallel to the fault and at some places more than 200 feet from it. The richest ore has been mined between level 3 and the level of Commodore adit No. 5. A set of stopes was raised in ore from the Nelson level to the level of tunnel 4, a distance on the incline of 515 feet, in a practically continuous ore shoot.

The distribution of the ore in depth, together with the abundance of oxidation products, including native silver, indicates that the ores have been enriched by secondary processes. The ore is not rich nor the ore bodies persistent along the Nelson adit, but at several places the lode is workable, and it is not certain that the bottom of the secondary ore has been reached. It is stated on good authority that ore carrying 45 ounces of silver to the ton was found in a winze sunk 125 feet below the Nelson tunnel.

COPPER LODE.

The Copper lode, which is on the Bachelor ground, is about 300 feet N. 57° W. of the portal of Commodore adit No. 3. A tunnel on the Copper claim is driven about 150 feet on a lode that strikes N. 68° W. and dips 75° SW. The lode is a fault fissure. The footwall crops out as a rugged cliff of the Willow Creek rhyolite, and the hanging wall, exposed in the tunnel, is the Campbell Mountain rhyolite. A shaft said to be 40 feet deep is sunk on the vein near the portal of the upper tunnel. The minerals include white and amethystine quartz, barite, limonite, wad, and copper carbonate. The ore is said to carry about 20 ounces of silver to the ton. Although the lode has produced a little ore, the operations on it have not been commercially successful.

A lode corresponding in position and attitude to the Copper lode has been developed in a crosscut run 180 feet westward from a point 145 feet in from the portal of the No. 3 tunnel of the Commodore.

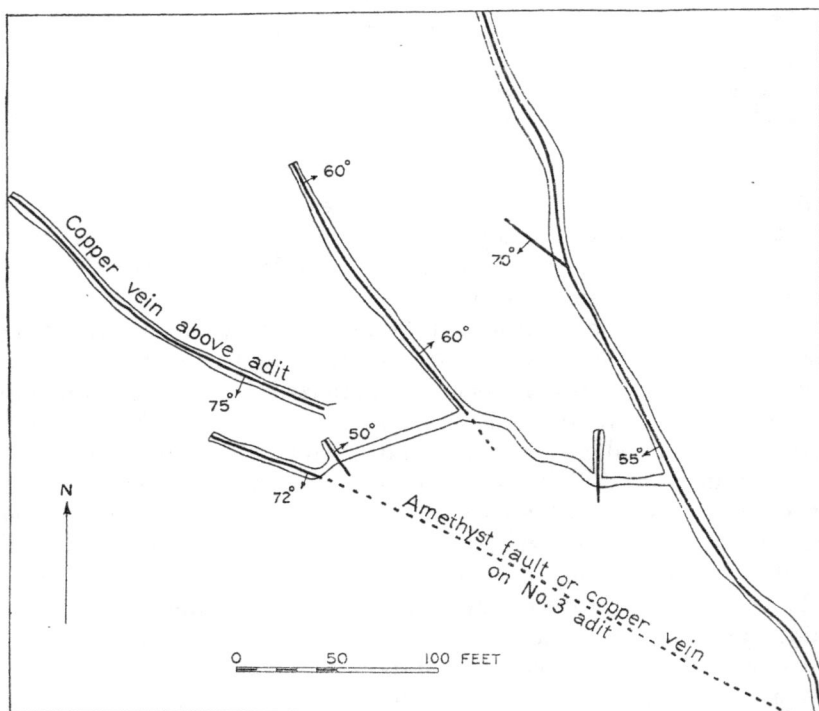

FIGURE 19.—Sketch of part of No. 3 adit, Bachelor mine, showing position of Amethyst fault on adit level and on Copper lode adit above level. Based on pace and compass survey.

It strikes N. 70° W. and dips 72° SW., having approximately the attitude of the Copper lode at the surface. As shown by figure 19, the block of ground between the Copper lode and the Commodore adit is highly fractured. Some of the fractures carry a little ore, but none are of stoping width.

COMMODORE MINE.

The Commodore mine, on the Amethyst lode, joins the Bachelor mine on the north and in turn is joined on the north by the New York mine. The Commodore claim and the Archimedes, a small fraction lying between the Commodore and Bachelor claims, are owned by the Commodore Mining Co. and have nearly 1,600 feet of the apex of the Amethyst vein.

The Commodore claim was located in April, 1891, by J. C. MacKenzie and W. V. McGilliard. Soon afterward it was sold to A. E. Reynolds, of Denver, and associates, who still own and operate the mine. The following figures, showing the total weight and value of ore produced to June, 1912, are available through the courtesy of Mr. Reynolds:

Gross weight	pounds..	788,489,388
Water	do....	61,254,618
Net weight	do....	727,234,770
Total production		$6,601,544.35
Freight		1,336,331.07
Net value		5,265,213,28
Total silver	ounces..	16,051,864
Total gold	do....	447
Total lead	pounds..	3,401,996

The total product of the mine, 363,617 tons net, has a value of $18.15 a ton. These figures show also that the ore carried 44.2 ounces of silver to the ton.

The Commodore is worked through the tunnels that pass through the Bachelor. The lowest of these is No. 5, the Manhattan tunnel. No. 4 is 399 feet higher, No. 3 is 800 feet higher, and No. 1 is 1,129.6 feet higher than No. 5. The portal of No. 2 between No. 1 and No. 3 is closed with waste. From the level of No. 5 a winze, inclined about 60°, is put down in the footwall of the vein to a depth of about 480 feet on the incline. Numerous workings down the dip of the vein connect the levels. The largest of these workings is the Discovery shaft, which extends from the apex to level 5. The Nelson tunnel does not encounter the vein in the Commodore ground but is driven approximately parallel to the Amethyst fault in the hanging wall of the fault.

The outcrop of the vein on the Commodore claim is not conspicuous, being nearly everywhere concealed by talus. At the discovery shaft the vein is not now exposed, but 350 feet north of the shaft the stopes are caved for 75 feet along the outcrop, and in this cave the slickensided hanging wall dips 60° SW. The Amethyst vein in the Commodore mine is nearly everywhere in the plane of the Amethyst fault. The fault and the vein join near the south end of the Commodore. On the levels of tunnel 1 and of the Nelson tunnel the junction is approximately at the boundary between the Commodore and Bachelor. On tunnels 3 and 4 the intersection is about 200 feet north of the Bachelor end line. (See Pl. III and fig. 18.)

The mineralization in the Commodore mine is more nearly uniform than that in the Bachelor. The principal vein is in the Amethyst fault fissure, whereas in the Bachelor mine nearly all the ore has been taken from fractures in the footwall of the fault. In the Commodore mine the footwall of the Amethyst fault is everywhere the Willow Creek rhyolite. The hanging wall in the lower levels is the Campbell Mountain rhyolite. On level 3 a raise driven from a point about 2,400 feet from the portal encountered a tuff, probably from the upper member of the Creede formation. At the level of adit 3, however, the hanging wall of the Amethyst fault is everywhere the Campbell Mountain rhyolite. The contact between the Campbell Mountain rhyolite and the Creede formation is above level 3, but its exact position is not known.

On level 1, about 180 feet north of the discovery shaft, a crosscut is run southwest about 100 feet. In this crosscut, 60 feet from the main tunnel, is a slickensided fault plane along which gouge is de-

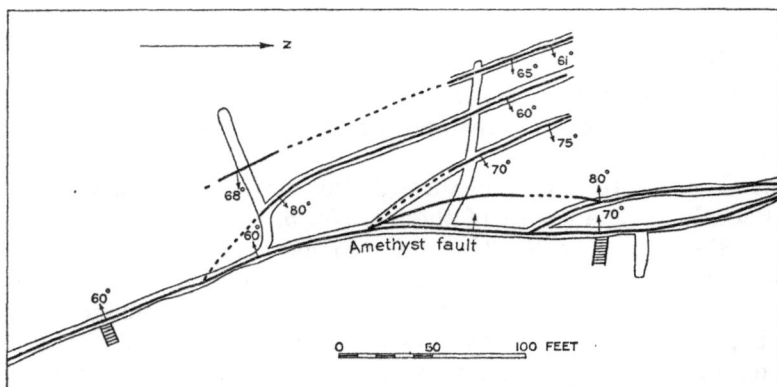

FIGURE 20.—Map of part of Commodore mine on first level below No. 1 adit. Shows parallel fractures in hanging wall of Amethyst fault. Based on pace and compass survey.

veloped. It strikes N. 26° W. and dips 50° W. On its footwall is a reddish phase of the upper member of the Creede formation, and on the hanging wall is a soft, porous, earthy red rock, probably a tuff belonging to the same formation. Ill-defined nearly horizontal planes probably indicate bedding. The formation carries fragments of rhyolite of many colors, and the matrix appears to be altered tuffaceous rhyolite.

About 375 feet north of the discovery shaft the drift turns sharply to the west. Here the hanging wall is of soft putty-like material of light-gray color, which carries snow-white masses as large as a hazel nut, probably altered fragments. At the point where the tunnel turns the fault shows the Willow Creek rhyolite in the footwall and the Creede formation in the hanging wall. The tunnel swings off on a fractured zone that makes an angle of about 35° with the Amethyst fault. The rock here is much fractured and altered. There are a few small rounded rhyolite fragments 4 to 6 inches in diameter in both the hanging wall and the footwall of the fault. This point is approximately in the line of a zone of highly fractured ground along the vein which extends to the surface and pitches southward about 70° in the plane of the vein.

Not all the ore of the Commodore mine is in the Amethyst fault fissure. On the level below adit No. 1 the fissuring in the hanging wall of the fault is complex. Four fissures nearly parallel with the fault have been developed, and locally these have supplied shipping ore. Figure 20 is based on a pace and compass sketch at this level, and figure 21 is a cross section of the vein in block 14 at a point 275 feet north of the discovery shaft.

The mineralization on the Amethyst fault in the Commodore mine is not uniform. There is very little workable ore above the level of tunnel 1, or within 200 feet of the surface. The most productive portion of the mine is between tunnels 1 and 3, but some very good ore has been mined between tunnels 3 and 4, and some stopes have been raised between tunnels 4 and 5. The great stopes between tunnels 1 and 3, from 200 to 500 feet below the surface, extended along the strike of the vein for nearly 1,500 feet. The ore at this level was of better grade and much more abundant than in the levels above and below. Most of the ore was oxidized, and the

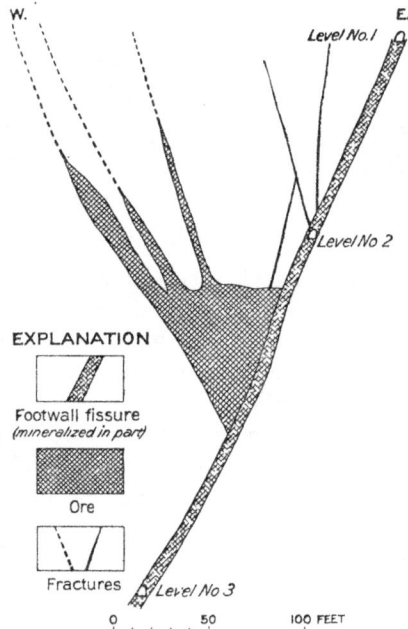

FIGURE 21.—Cross section on block 14, Commodore mine, showing fractures in hanging wall.

principal minerals were native silver, cerargyrite, pyromorphite, anglesite, cerusite, and smithsonite, with considerable limonite and wad. This ore carries very little sulphide. In depth galena, sphalerite, pyrite, and chalcopyrite become increasingly abundant. The gangue is mainly white and amethystine quartz and barite.

Postmineral fracturing is pronounced at all levels, and the ore as deep as level 5, 1,450 feet below the surface, is partly oxidized. In a stope about 150 feet north of the line of ore chutes which is termed the Discovery shaft, between 1,100 and 1,200 feet below the surface, wire silver, undoubtedly an alteration product, was found in considerable quantity. A fine specimen of ore from the upper levels of the Commodore mine, in the possession of Mr. W. G. Messinger, of Creede, shows gray quartz, with galena, zinc blende, and a little limonite and manganese oxide. Woolly wads of hair-like wire silver occur in vugs in the quartz. Some polished specimens owned by Mr. J. F. Wilson, foreman of the mine, consist of alternating bands of red jasper, white quartz, and leaf silver.

The following partial analyses of the ore are supplied by courtesy of the American Smelting & Refining Co.

Partial analyses of ore from Commodore mine.

[Per cent except as otherwise indicated.]

Date.	Tons.	Ag (ounces to the ton).	Pb.	SiO$_2$.	Fe, Mn.	Al$_2$O$_3$.	Zn.	CaO.	BaSO$_4$.	
December, 1903	1,450	32.5	2.8	65.5	4	6.0	0.6	0.3	9.5	1.0
December, 1904	1,339	32.7	2.9	59.6	6.9	3.2	.7	.4	13.8	1.7
March, 1905	1,293	32.7	1.5	62.5	6.2	3.4	.7	.6	15.5	1.9
February, 1906	807	23.6	1.5	65.1	6.0	5.4	1.4	.6	7.5	1.7
December, 1906	2,006	20.7	1.4	66.7	3.9	5.7	.6		8.7	2.2

LAST CHANCE, NEW YORK, AND DEL MONTE MINES.

The Last Chance, New York, and Del Monte claims overlap one another in such a manner that the total rights of the three claims cover about those of a full-length claim. They lie between the Commodore mine on the south and the Amethyst mine on the north. The three claims are now leased to one company and worked from a single shaft on the Last Chance claim. They will be described as one mine, although they are not held by precisely the same owners.

The Last Chance claim was located in August, 1891, by Theodore Renniger and Julius Haas, who were grubstaked by Ralph Granger and Eric Buddenbock, of Del Monte. Although the location followed Creede's discovery of the deposit at the Holy Moses mine, it was the discovery of the rich ore in the Last Chance that resulted in the rapid development of the district. The Last Chance passed ultimately into the hands of the Last Chance Mining & Milling Co., owned principally by the E. O. Walcott estate, Jake Saunders, and John Morgan, of Denver. The New York was located by George K. Smith and S. D. Coffin. Soon afterward Coffin sold his interests to A. E. Reynolds, I. L. Johnson, and associates, of Denver. These two groups of owners for a long period were engaged in a legal dispute over apex rights. Finally a compromise dividing the disputed ground was effected.

The Volunteer Mining Co., which has large interests in this group, is a holding company which owns stock of the New York and also of the Last Chance Co. The three claims are now under lease to the Del Monte Leasing Co. At present the lease is operated by A. E. Humphreys, of Denver, and Albert Collins, of Creede. The ownership of this group of mines is complexly involved, and records for the entire period of production are not available to the writers. Mr. Albert Collins estimates the production of the group at $19,000,000, the larger part of which he credits to the Last Chance. The portion of the vein included in this group was much richer than any other section of an equal number of linear feet on the Amethyst lode.

The New York and Del Monte shafts are not accessible. The Last Chance, New York, and Del Monte claims are worked through the Last Chance shaft and the Nelson tunnel. The shaft is about 1,400 feet long and is sunk on or near the vein for the whole of its length. Near the surface it is inclined about 70°, but within 150 feet of the surface it flattens to 55°, and below this point its inclination is from 55° to 50°, for the most part about 51°. The shaft does not follow down the dip of the vein, but makes an angle with the dip of nearly 30° N. Its depth on level 12 is about 1,112 feet vertically below the collar. This level is 37 feet vertically above the Nelson tunnel, making the total depth of the mine about 1,150 feet below the collar of the shaft. Twelve levels (1½ to 12) are turned from the shaft at intervals of about 100 or 200 feet on the incline. Above the collar of the shaft a short adit is driven southward on the vein. There are altogether over 2 miles of drifts and crosscuts.

The outcrop of the lode is here about 10,200 feet above sea level. It is everywhere oxidized, but, unlike the lode at the Commodore and Bachelor mines, the outcrop is not everywhere leached and unworkable. Near the Last Chance shaft the vein was stoped at the surface, and according to report some of the ore taken within a few feet of the surface was above the average in value. On the Amethyst and Last Chance ground the outcrop of the lode has not suffered glaciation, the edge of the West Willow Creek glacier having halted higher up the canyon.

The lode in the Last Chance and New York mines lies in and along the Amethyst fault. The footwall of the fault wherever exposed is the Willow Creek rhyolite. On the section down the Last Chance shaft (fig. 23) the distinctions between the Willow Creek rhyolite and the Windy Gulch rhyolite breccia, which forms the hanging wall of the Amethyst fault near level 7, are clearly shown. Below level 7 the hanging wall of the fault is the Campbell Mountain rhyolite, and the contrasts between this rock and the Willow Creek rhyolite are at most places clear. Along the shaft, however, between levels 7 and 11, there is a thin slab of rhyolite that is probably less than 20 feet thick at most places. It is pinkish purple or light red in color but shows very little of the red mottling characteristic of the Campbell Mountain rhyolite. At one place near the shaft on level 7 mottled red rhyolite and gray rhyolite grade into each other. On level 10, where the station is cut out near the shaft, a tight fault fissure nearly parallel to the vein is exposed. The hanging wall of the fissure is typical mottled red rhyolite of the Campbell Mountain formation. The footwall is gray rhyolite and resembles the Willow Creek rhyolite, but in view of the fact that a gray phase is developed locally in the Campbell Mountain rhyolite it is not certain that this slab is the Willow Creek rhyolite. If it is, the Amethyst vein along the shaft from level 7 to level 11 lies a few feet in the footwall of the main fault. These relations are indicated in figure 23, in which the formation that constitutes this particular block of ground is not indicated.

As stated above, the red Campbell Mountain rhyolite forms the hanging wall in the lower levels, extending from the adit level up nearly to level 7. On level 6 and from that level nearly to the surface the hanging wall of the fault is the Windy Gulch rhyolite breccia.

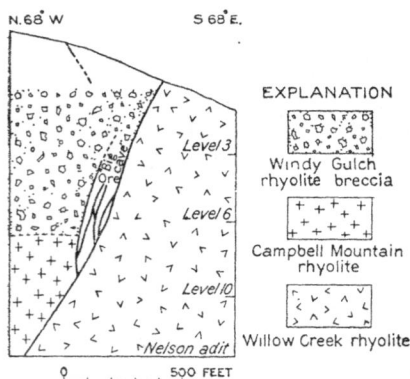

FIGURE 22.—Cross section through Amethyst lode 150 feet south of Last Chance shaft. Direction of section, S. 60° W.

This breccia in the Last Chance workings rests directly upon the Campbell Mountain rhyolite, and as both these formations are rhyolite breccias, it is not everywhere possible to distinguish one from the other, and the contact is drawn somewhat arbitrarily. It is possible that the division may lie between levels 7 and 8 of the Last Chance rather than between levels 6 and 7, as shown in figures 22 and 23.

On the tunnel level (Last Chance 13) an 80-foot crosscut is run from the tunnel to the bottom of the Last Chance shaft. At 60 feet from the main tunnel it cuts a vein that strikes N. 20° W. and dips 50° W. This is probably the Amethyst vein, but it is not developed on this level. The hanging wall is the hornblende-quartz latite porphyry, probably an intrusive dike. In the crosscut to the shaft it is about 20 feet wide. A similar rock is exposed about 150 feet north of the point where the crosscut joins the main tunnel. On level 12 of the Last Chance mine this porphyry was noted in the hanging wall at the shaft, and it extends southward from the shaft 16 feet or more.

The lode is a sheeted zone. The most persistent fissure is on the footwall side of the lode, which dips in general 55°–70° W., prevailingly at the lower angles. Where crosscuts have been run to the west of the fissures, veins have generally been encountered in the hanging wall. Many of them have been stoped, but some are too narrow or too low in grade to be profitably worked. On level 6 there are six veins, five of which were worked for ore (fig. 22). The country rock between these veins is silicified and sericitized and at some places is said to carry a little silver. The dip of the vein appears to be lower on the section shown in figure 23 than on that shown in figure 22. This is due to the fact that the section of figure 22 is approximately down the dip, whereas the section of figure 23 is along the shaft, which makes an angle of about 30° with the steepest dip.

On level 6 the mineralized zone at the widest place is 100 feet wide, the workable ore being confined, however, to the six fissures above noted. On levels 3, 4, and 5 there was a great body of ore known as the "big cave." This ore was so much fractured and the ground so unstable that it was mined by milling the dirt from the bottom on levels 5 and 6 and drawing it out by gravity from levels as high as No. 2. It is said that many thousand tons of ore was thus mined and, with little sorting, sent to the smelters. The cave left by drawing out the ore is inaccessible but is credibly reported to be in places about 100 feet wide. The stopes are elsewhere about 3 to 5 feet wide, and the great width of ore in the cave is probably due to the close spacing of parallel veins.

The Amethyst fault is the most persistent fissure. Except from the "big cave" the hanging-wall fissures have supplied comparatively

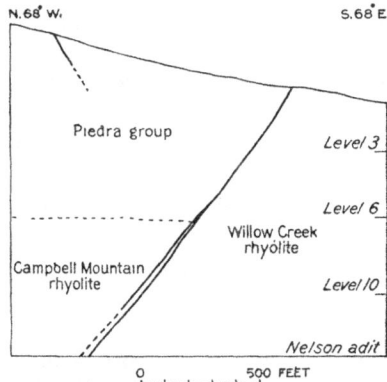

FIGURE 23.—Cross section of Amethyst lode along Last Chance shaft. Direction of section, N. 68° W.

little ore. The continuity of the ore along the fault fissure and its regularity, especially in the upper levels, are noteworthy. There are few places in the developed portion of the mine where the foot-wall vein has not been stoped. The hanging-wall veins of the lode are less persistent, generally less valuable, and not so wide. Some of them die out toward the surface, and practically all of them join the main vein in depth.

Most of the subordinate fissures in the Last Chance and New York mines dip west, like the principal vein, but as they dip at higher angles they join the lode in depth. In the Commodore mine the hanging-wall fractures join the main vein at larger angles.

The Del Monte vein, which is represented in figure 22 as dipping 60° toward the Amethyst fault, is said to join it in the "big cave." This vein has produced considerable ore; it was not accessible in 1912.

The 13 levels that are developed in the Last Chance mine are driven in the main on the vein of the Amethyst fault, but on levels 3 to 8, south of the shaft, drifts are run on veins in the hanging wall.

Level 1 is an adit, whose portal was about 60 feet south of the shaft. There the vein was stoped at the surface, and a great pit is now exposed, showing on the walls the Windy Gulch rhyolite breccia. The portal of the adit is caved.

Level 2, which is about 117 feet below the collar of the shaft, is driven on the Amethyst fault. About 250 feet south of the shaft this level is not accessible.

Level 3 was driven 920 feet south from the shaft but is caved 625 feet south of the shaft. The accessible part of this level is driven on the Amethyst fault.

Level 4 is caved 200 feet south of the shaft.

Level 5 follows the Amethyst fault for 200 feet south of the shaft to the "big cave." About 110 feet south of the shaft a crosscut is run 200 feet in the Willow Creek rhyolite. At 290 feet from the shaft the crosscut makes a right angle turn toward the west, and at 80 feet beyond the turn it encounters the vein. The drift is caved a few feet south of this point, but the map of the mine shows that it is run nearly 800 feet farther south.

On level 6 the developments are more extensive than on the other levels. About 500 feet south of the shaft six veins are developed. Of these one is in the fault fissure, one is near and nearly parallel to it in the footwall, and four are in the hanging wall. Here the fractured zone is about 100 feet wide, and much of it is silicified and mineralized. The cross section (fig. 22) shows only six of these fissures, but there are probably subordinate ones, especially in the cave about level 6, which was mined by drawing ore and rock without blasting.

Level 7, at the shaft, is driven on the vein. About 250 feet south of the shaft a small cave a foot or two in the west wall exposes typical mottled red rhyolite belonging to the Campbell Mountain rhyolite. About 350 feet south of the shaft the vein is wide. The main drift runs in the footwall, and 20 feet west of it a drift runs south, parallel to it. This drift follows a vein that occupies for the most part the fault between the Willow Creek rhyolite and the Campbell Mountain rhyolite. The fault plane dips about 65° W. Between the footwall of this fissure and the main drift is a block of country rock 20 feet wide, which is highly shattered and mineralized, constituting a low-grade ore. Large fragments of the Willow Creek and Campbell Mountain rhyolites are found in it, together with much quartz and white clay. This level is driven about 1,100 feet south of the shaft, but the south end is not accessible.

On level 8 the stoped vein lies below and about 10 feet east of the shaft. The hanging wall of the vein for about 160 feet north and 140 feet south from the shaft is reddish-purple rhyolite. This rock may be a thin slice of the Willow Creek rhyolite, not much brecciated, lying between the vein and the fault. The level is run 1,000 feet south of the shaft, almost to the end line, but is caved 600 feet south of the shaft.

On level 9, from the shaft to a cave 175 feet south of the shaft, the footwall is the Willow Creek rhyolite and the hanging wall is purplish rhyolite like that above the shaft on the higher levels.

On level 10 the shaft is driven on the vein. The footwall is the Willow Creek rhyolite. About 12 feet west of the vein, where the station for the shaft is cut, a fault is exposed for 22 feet along the strike. Here the typical Campbell Mountain mottled red rhyolite is shown in the hanging wall of the fissure. The footwall contains only a few fragments and may be the Willow Creek rhyolite. The fault and the vein that has been developed in the footwall probably join not more than 125 feet south of the shaft on level 8, for red Campbell rhyolite is exposed in the hanging wall at that point.

Level 11, driven on the vein, is caved 250 feet south of the shaft.

The ore is worked for silver, lead, and gold. The metallic minerals are galena, zinc blende, and pyrite. The gangue is mainly quartz and barite. The oxidation products include limonite, hematite, malachite, anglesite, cerusite, goslarite, and manganese oxide. In the oxidized ore the valuable constituents appear to be mainly native silver and silver chloride, cerusite, and anglesite. The quartz includes white, blue, and amethystine varieties. Barite is abundant in the ore along the outcrop and as deep as level 3, more than 300 feet below the surface. It is probably present also as low as level 6, but it is much less abundant on the deeper levels than near the surface. Oxidation is thorough in the upper levels, and the ore is partly oxidized down to level 12 inclusive. In levels 10 to 12 it consists of the sulphides with considerable quartz cut by numerous closely and irregularly spaced fractures.

About 175 feet south of the shaft, in the stope above level 11, the vein, as shown in figure 7, is 3 feet wide and dips 58° W. The ore consists of quartz, galena, and zinc blende and is said to carry 18 ounces of silver and about $2 in gold to the ton and 8 per cent of lead. The vein is inclosed in rhyolite flow breccia and includes fragments of the breccia over a foot long and 4 or 5 inches wide. A 3-inch veinlet composed almost entirely of galena and zinc blende cuts across the vein nearly at right angles to the wall. This veinlet is in turn cut by a second veinlet of white barren quartz, which is about 2 inches wide and extends across the principal vein. The vein of quartz is strikingly banded and contains a vug about 6 inches long and an inch wide lined with crystals of white quartz projecting toward the center. The relations are indicated in figure 7 (p. 114).

About 200 feet below this point on the dip of the vein, on level 12, the vein dips 55° W. It consists of a highly crushed zone 6 feet wide inclosed in rhyolite breccia. The hanging wall is a well-defined slickensided plane surface, and the footwall is somewhat undulatory. On the footwall side is about 4 feet of clay gouge containing numerous fragments of altered rhyolite and breccia. This mass is so thoroughly crushed that it is practically a mass of fine mud containing silicified fragments of the country rock. It is said to carry about 8 ounces of silver to the ton. Above this low-grade material is about 2 feet or more of highly siliceous ore, consisting of blue quartz, galena, and zinc blende, stained with iron and manganese oxides, which is said to carry 18 ounces of silver and $2 in gold to the ton and 9 per cent of lead. This ore is crushed but is more siliceous than the low-grade material below it, and is not so claylike in consistency. No barite was noted in this ore.

In the table below are analyses of ore taken from the Last Chance mine by the Del Monte Leasing Co. Nos. 1 and 2 are "high-grade" ore, the others are "low-grade" ore.

Analyses of ore from Last Chance mine.

[Per cent except as otherwise indicated.]

	Date.	Tons.	Au (ounces to the ton).	Ag (ounces to the ton).	Pb.	SiO₂.	Fe,Mn.	Al₂O₃.	Zn.	CaO.	BaSO₄.	S.
1	September, 1903.........	174	0.04	40.1	3.7	69.3	3.4	6.6	0.4	1.	....	1.8
2	February, 1904	392	.04	66.1	3.1	68.1	4.	6.5	......	......	1.9	1.1
3	May, 1904.....	1,785	.08	21.0	4.2	74.2	2.8	4.6	2.	......	Tr.	1.6
4	March, 1905...	1,814	.05	25.5	3.9	68.4	7.3	5.4	......	0.7	....	1.2
5	February, 1906	1,406	.03	24.9	2.9	67.6	8.	5.2	......	1.	....	1.3

The ore has been greatly crushed since it was deposited, and descending waters have had free access to it. Some oxidation has taken place along fractures as deep as the lowest level, but the ore is not thoroughly oxidized below level 5. The richest ore and the bulk of the shipping ore has been obtained above level 7 about 700 feet below the surface. Nearly all of it shows considerable oxidation. These relations indicate that the richer ore bodies are due in part to processes of sulphide enrichment. The ore mined near the surface carried very little gold, but the solid sulphide ore from the lower levels carries in general from $1 to $4 in gold to the ton.

AMETHYST MINE.

The Amethyst mine is on the steep slope of Bachelor Mountain, on the west side of West Willow Creek, between the Last Chance and Happy Thought mines. The claim was located in August, 1891, by N. C. Creede, D. H. Moffat, and L. E. Campbell. Amethystine quartz, rich in gold and silver, was found practically at the outcrop, and from this the claim received its name. The mine was one of the first in the district to produce ore, and with the exception of a few relatively short intervals when it was shut down, it has been producing steadily since. The total production of ore and concentrates is estimated to be about $4,000,000. In 1911 the mine was closed, but in 1912 it was under lease to the Creede Triune Mining Co. This company installed an electric power plant at Creede to effect a saving in freighting coal by wagon road. A three-compartment blind shaft on the Nelson adit level was sunk on the vein near the north end of the mine. Some new stopes were opened, and the mill was put into operation.

The mine was worked through the Nelson tunnel and the Amethyst shaft. The shaft is about 1,283 feet long, as measured on the incline, and is sunk on the vein, which dips about 55° W. The shaft does not follow exactly the line of steepest dip but is inclined to the south in the plane of the vein at an angle 3° or 4° lower than the dip of the vein. Twelve levels are turned from the shaft at intervals of about 100 to 200 feet on the incline. They range in length from 800 to 1,400 feet. The total underground workings, mainly drifts on the Amethyst vein, cover more than 2½ miles. An adit 870 feet long driven westward from West Willow Creek connects with level 5 about 350 feet vertically below the collar of the Amethyst shaft. On level 2 a drift 300 feet long is run into the hanging wall from the shaft. It is reported to be driven on a thin fissure. In 1912 this drift was caved. On level 12, about 570 feet north of the shaft, a crosscut about 200 feet long is driven in the hanging wall. Except on levels 2 and 12, no work has been done in the hanging wall at the vein, all the levels being run along it.

In earlier years, when the rich oxidized ores of the upper levels were mined, the ore was loaded into a tramcar at the shaft house and sent over an aerial tramway to North Creede, where it was loaded into railway cars and sent to smelters. This tramway is now in ruins, and a shorter one connects with a mill on West Willow Creek about 800 feet from the Amethyst shaft. This mill is equipped with crushers, rolls, trommels, jigs, tables, and canvas plant and is said to give a satisfactory saving when the sulphide ores are treated.

The lode crops out at several places between the Last Chance and Amethyst shafts, and for a considerable part of this interval the stopes are raised to the surface. No ore was seen at the surface, but the wall rock along the outcrop is stained here and there with limonite and carries veinlets of quartz, barite, and jarosite.

The lode, as developed in the Amethyst mine, is on the Amethyst fault. The footwall from the surface to level 13 (the tunnel level) is the Willow Creek rhyolite. Although there are some outcrops of this formation on the surface east of the Amethyst vein, the best exposure in the vicinity of the Amethyst mine is in the crosscut on level 5 which is run entirely in this formation, from West Willow Creek to the vein. In this adit the formation is mainly red rhyolite streaked with white, but in places it is purplish, chocolate-colored, or gray. The footwall is red also on level 6, but on level 7 the grayish and grayish-green phases predominate, as they do on the lower levels and in the Nelson adit. The hanging wall at the surface where exposed in a caved stope just above the shaft house is the Windy Gulch rhyolite breccia. Not much of the mine is open above level 5, but a somewhat difficult passage may be made through partly caved stopes along and near an old incline about 300 feet south of the main shaft. Exposures of the wall rock are not numerous, and the rock and oxidized vein stuff are so highly altered that the determination of the formation is not easy, but no typical red Campbell Mountain rhyolite was noted on or above level 4. The contact between the Campbell Mountain rhyolite and the Windy Gulch breccia is therefore drawn between levels 4 and 5. A cross section of the mine is shown in figure 24, and one of the vein is shown in figure 25.

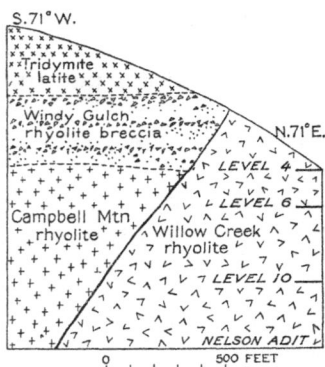

FIGURE 24.—Cross section of Amethyst mine 175 feet south of shaft.

From level 5 downward nearly everywhere a difference is shown between the rhyolite of the footwall and that of the hanging wall, the vein being clearly in the fault fissure. On level 5 in the hanging wall the common phase, or mottled red rhyolite, prevails. On level 6 there is not much mottling and very few fragments are present. This phase prevails in the hanging wall on levels 7 and 9 also, certain parts of the hanging wall having much of the same appearance as the red or gray rhyolite of the Willow Creek rhyolite. On levels 10 and 11 the hanging wall is mottled, as it is on level 5. On level 12 and on the tunnel level some red phases are shown, but gray phases predominate, although the rhyolite is mottled much like the typical red Campbell Mountain rhyolite of higher levels. The typical red mottling is shown on level 11, but the grayish mottled breccia on level 12 and on the tunnel level may possibly he a transitional phase between the Campbell Mountain rhyolite and the Willow Creek rhyolite.

The most productive ore bodies of the Amethyst mine were found above level 4 and within 300 feet of the surface. Stopes were practically continuous from the end line of the Last Chance and extended to or nearly to the surface. Below level 4 the shoot of better-grade ore extended downward south of the shaft, where it pitched somewhat to the south and also north of the shaft, where it pitched somewhat to the north. The shaft itself below level 4 was driven in a relatively poor portion of the vein.

The minerals of the ore include quartz, chlorite, barite, galena, sphalerite, pyrite, chalcopyrite, cerusite, anglesite, limonite, goslarite, wad, native silver, and gold. Cerargyrite and pyromorphite were noted in cabinet specimens from the upper levels. To judge from the dump, amethystine quartz is more abundant here than on this lode in the mines to the south, and green chlorite appears in great abundance, especially on level 12, where it is the chief gangue mineral in some of the ore.

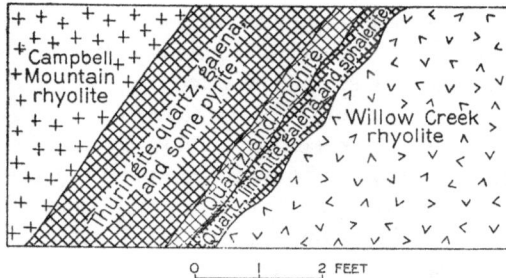

FIGURE 25.—Section of Amethyst vein, Amethyst mine, in stopes above level 9, about 120 feet north of Amethyst shaft.

This mineral is not conspicuous in the Commodore, Last Chance, and Bachelor mines.

The development of sulphates is noteworthy. Iron sulphates incrust the walls, and goslarite was noted at many places. The hairlike wads of this mineral form thick nests a foot or more in diameter that were evidently deposited by trickling waters carrying zinc sulphate long after the mine was opened. Some of them are forming now. Barite and jarosite were noted along the outcrop. Barite was doubtless abundant also in the upper levels, but these are accessible now at only a few places, and most of the ore has been removed. On levels 5 and 6 barite is conspicuously developed, especially along seams and fractures in the altered ore. On these levels, 250 feet north of the shaft, the ore is almost completely oxidized and is said to carry 5 to 28 ounces of silver to the ton for a width of 20 feet. Much of this ore is cut by barite seams, which follow small fractures in the rock. Broken pieces of the altered rock and ore are coated over with thinly spaced crystals of barite about 0.1 inch wide and 0.2 to 1.5 inches long. Similar barite crystals lie directly below this point on level 6 in similar oxidized ore.   There is a great caved stope at this place on level 5.   It is about 20 feet wide, and all of it is in low-grade mineralized material.   On level 7 barite is less abundant.

On and above level 4 oxidation is nearly complete, but here and there small nodules of galena remain as cores in oxidized material. This ore, it is said, carried from 50 to 100 ounces to the ton in silver, with a small percentage of lead and only a little gold.   Oxidation is almost complete also on level 5 and north of the shaft on level 6. South of the shaft on level 6 and on the lower levels the oxidation is confined to fractures and thin seams that cut the sulphide ore.   In the lower levels the ore carries in general about 5 to 10 per cent of lead, with 8 to 10 ounces of silver and $1 to $2 in gold to the ton. In a stope above level 9, about 120 feet north of the shaft, the ore body is 6 feet wide and dips 54° W.   It carries 16 per cent of lead as galena and 4 per cent of zinc as sphalerite.

In a small stope just below level 12, about 800 feet north of the shaft, the ore consists of galena and zinc blende in a gangue of quartz and chloride, stained with copper carbonate and iron and manganese oxides.   It is said to carry $5 to $15 in gold to the ton, with some silver and lead.   A specimen owned by Mr. W. G. Messinger, of Creede, reported to have come from level 12 of the Amethyst, probably from the gold stope mentioned above, consists of blue-gray quartz, with some amethystine quartz, green chlorite, galena, and zinc blende. Thin fissures of lead carbonate cross this ore.   Fine wire gold of deep-yellow color occurs in small cavities.   The rock carries a little limonite, and dark powdery manganese oxide is associated with some of the gold.   Some of it, however, appears to be free from manganese.

On level 13, about 570 feet north of the shaft and 390 feet south of the three-compartment winze or blind shaft sunk from this level, a vein has been developed in the hanging wall.   It strikes N. 4° W., making an angle about 15° with the main vein, the intersection pointing north.   It dips 45° W.   Both walls are in the Campbell Mountain rhyolite and in the hanging wall of the main fault.   This vein contains much green chlorite, sphalerite, and galena, with a fair amount of silver.   Very little barite was noted.   The vein is about 2 feet wide and carries concentrating ore of good grade, which yields mainly silver and lead.

The main vein, or fault fissure, strikes about N. 20° W. and dips at most places 55°–58° SW.   It ranges in width from 1 to 12 feet but in general is from 3 to 6 feet wide.   Everywhere it is fractured, and in some places it is highly crushed.   Almost invariably a slickensided surface marks the hanging wall or footwall, and at some places both walls are slickensided.   At many places the slickensided surfaces are grooved and striated.   On an intermediate level, about 50 feet above level 4, near an incline, now abandoned, that was put down 300 feet south of the collar of the main shaft, the hanging wall dips 55° W. Striae making an angle of 45° with the line of steepest dip pitch south. Another example of pitching striae was noted in the stopes above level 7, about 150 feet north of the main shaft. At this place the slickensided plane cuts through the ore, and the dark sulphides, mainly pyrite and galena, are polished smooth. Deep striae plunge about 85° N., and only 10 feet away striae plunging 85° S. were noted. On level 9, 50 feet south of the shaft, a slickensided surface shows striae pitching 85° S. In the stopes above level 11, about 100 feet south of the shaft, a slickensided surface shows striae pitching 75° N., and about 10 feet south of this point the slickensided surface shows striae pitching 75° S. The striae seem to indicate that the movement, so far as the horizontal element is concerned, was to a certain extent compensating. The movement of the hanging wall was at one time to the north, at another to the south—a zigzag motion, with the main element downward.

The lode in the Amethyst mine shows few noteworthy features that have not already been mentioned in the description of mines on the Amethyst lode farther south. Practically all the ore mined has been taken from the Amethyst fault fissure. Aside from the two short drifts that have been run on levels 2 and 12, the hanging wall is unexplored. Possibly later work will show that the fracturing in the hanging wall of the Amethyst mine is as complex as it is in the Commodore and Last Chance mines. The remarkable development of a green ferruginous chlorite, the abundance of amethystine quartz, and the distribution of sulphates in depth have already been mentioned. The fractured condition of the lode, the distribution of the valuable metals with respect to depth, and the paragenetic evidence point clearly to a rearrangement of the metals in the lode by sulphide enrichment.

HAPPY THOUGHT MINE.

The Happy Thought mine is on the Amethyst vein north of the Amethyst mine and south of the White Star claim, which lies between the Happy Thought and Park Regent mines. The Happy Thought is owned by the Creede United Mines Co., of Denver. It is leased by the North Creede Mining Co., which has recently subleased it to the Creede Local Mining Co.

The Happy Thought, with the Argenta, which lies just west of it, carries about 1,600 feet of the apex of the Amethyst lode. The mine is worked through a shaft sunk down the dip of the vein at an angle of 55°. The shaft is about 1,400 feet long as measured on the incline and connects with the Humphreys tunnel, an extension of the Nelson tunnel. The vertical depth of the shaft is 1,182 feet. Levels are turned from the shaft at intervals of about 100 feet.

Survey stations 25 feet apart are numbered consecutively north and south of the shaft.

In December, 1907, a disastrous fire destroyed the shaft house, burned the timbers in the upper portion of the mine, and wrecked the shaft.   Only the lower part of the mine, including levels 7 to 12 has been reopened.   The shaft is not now in use, the Humphreys tunnel serving as the outlet.   The upper part of the mine, from the surface to a depth of about 600 feet, is not accessible.   On level 5 a drift is run from the shaft to a point near the south end line, where it connects with the surface.   Three short crosscuts are run into the hanging wall, one on level 7, one on level 8, and one on level 11.

Accurate figures for the production of the mine are not available. Mr. Albert Collins, who has been in close touch with the operations for many years, estimates the total production at $3,000,000.   This amount is smaller than the yields from the Amethyst, Last Chance, and Commodore mines, but it is probably greater than the production of either of these mines from levels more than 800 feet below the outcrop.   The production from the mines of the Amethyst lode south of the Happy Thought has come mainly from the upper workings, but in the Happy Thought the lower workings, from 600 to 1,200 feet below the surface, have been the most productive.   The ratio of gold to silver output is much higher in the Happy Thought mine than in the mines on the Amethyst lode south of the Happy Thought.

The Amethyst vein in the Happy Thought mine is in the Amethyst fault fissure.   The outcrop is concealed by surface débris, but in the mine the fault strikes about N. 25° W. and dips about 55° W. The footwall at most places is the Willow Creek rhyolite, but on levels 10 to 13 an altered hornblende-quartz latite porphyry, probably a dike, is found in places on the footwall.   On level 10 this porphyry is exposed at a point 488 feet north of the shaft and extends southward 208 feet.   On level 11 the altered porphyry forms the footwall 300 feet north of the shaft and extends southward to a point 150 feet north of the shaft.   From this point the footwall is concealed to station 4 north, or 100 feet north of the shaft.   From that station to station 2 north, or 50 feet north of the shaft, the Willow Creek rhyolite is exposed.   The porphyry reappears 50 feet north of the shaft and extends southward 100 feet.   Farther south to the end line, the footwall is the Willow Creek rhyolite.   A section showing the vein on level 10 is given in figure 26.

On level 12 the porphyry is exposed about 625 feet north of the shaft and extends northward on the footwall 25 feet.   South of this body the footwall is the banded Willow Creek rhyolite.   This is probably the same body of porphyry shown 600 feet north of the Happy Thought shaft on the geologic map of the Nelson adit level

(Pl. XII). The hanging wall of the Amethyst fault, wherever it has been observed in the Happy Thought mine, is the Campbell Mountain rhyolite. The Windy Gulch rhyolite breccia, which is exposed at the surface on the hanging wall of the fault at the Amethyst mine, probably forms the hanging wall in the upper part of the Happy Thought also, but if so it is concealed by surface débris. This rhyolite may have been encountered on the hanging wall in the upper workings, but those above level 7 are not accessible. Except on the tunnel level, there are no workings in the footwall of the lode. The relations between the Willow Creek rhyolite and the altered porphyry are generally not shown. Nearly everywhere the points of contact are lagged or otherwise concealed. On level 11, however, about 50 feet north of the shaft, the contact of the two formations is well exposed. The contact here shows some movement. It strikes N. 10° W. and dips 45° W. The Willow Creek rhyolite forms the hang ing wall. On the tunnel level, about 600 feet north of the shaft, the contact is crosscutting. These relations, together with the distribution of the porphyry else-where in the region, are in accord with the con-clusion that it is a dike intruded into the Willow

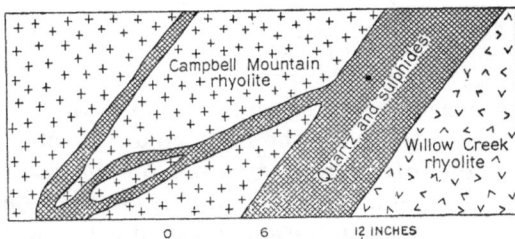

FIGURE 26.—Section showing vein in fault on level 10, Happy Thought mine, 375 feet north of shaft.

Creek rhyolite. At the Last Chance shaft the porphyry is on the hanging wall of the fault and intruded into the Campbell Mountain rhyolite. The porphyry is therefore younger than the Campbell Mountain rhyolite but older than the last period of movement on the Amethyst fault. It does not appear probable that the porphyry on the footwall of the Amethyst fault in the Happy Thought mine and that on the hanging wall of the Last Chance shaft are faulted portions of the same body. These exposures are about 1,700 feet apart, and no small masses of crushed porphyry were noted in the Amethyst mine, which lies between these bodies.

The vein is in general from 1 to 10 feet wide. At many places one or both walls are smooth or slickensided, and at some places the striae are nearly vertical. On level 10, about 450 feet south of the shaft, the striae on the footwall pitch 70° S.

The minerals include galena, sphalerite, pyrite, chalcopyrite, cerusite, anglesite, limonite, wad, native gold, and silver. The quartz includes white, gray, and amethystine varieties. Green chlorite is fairly abundant. Barite is less common than in the mines to the south, silver is much less common, and gold is more abundant. In the unoxidized ore of the lower levels green chlorite is an abundant gangue mineral. Some of the ore consists mainly of this chlorite, heavily impregnated with sphalerite and galena and carrying small proportions of gold and silver. This ore is much like that of the Solomon and Ridge mines, on the Solomon vein.

In the table below are given assays of concentrates of ore from the Happy Thought mine, supplied through the courtesy of Mr. A. E. Humphreys, president of the Creede United Mines Co.

Assays of concentrates from Happy Thought mine.

Tons.	Nature.	Zinc (per cent).	Lead (per cent).	Silver (ounces to the ton).	Gold (ounces to the ton).
877.1	Zinc concentrates.....	37	8.10	4.5	0.4
405.07	.....do..............	25–31	7–11	2–3	.32–.52
1,615.4	.....do..............	28	10	5	.3
426.4	Lead.................	(?)	65	6.9	1.1
325.9	.....do..............	(?)	52	6	1
517.2	.....do..............	....	68	5.5	1.3
49.4	Slimes...............	....	21.7	4.6	.48

A survey of these and of many other assays shows that the lead concentrates carry the most gold. Much of the gold is free, and in the concentration of this ore on the tables, where the rich ore is milled, a yellow line of gold particles may be observed above the lead. In the separation the gold is saved with galena, anglesite, and cerusite. Several tests were made to ascertain the association of the finer particles of gold. The concentrates consisted of anglesite and galena. In a gold pan, in the angle between the bottom and the side, the larger gold particles separate easily from the anglesite and galena. These tests showed that much of the gold which is separated with the lead minerals occurs in the ore as minute yellow particles associated with iron and manganese oxides. A smaller amount may be intimately associated with lead and zinc minerals.

The ore is highly oxidized to a depth of about 700 feet. South of the shaft the surface is somewhat lower than it is north of the shaft, and the oxidation has advanced to greater depths. The line of almost complete oxidation is 50 feet below level 9 south of the shaft and perhaps 20 feet above level 9 north of the shaft. On the tenth and lower levels the ore is mainly unoxidized, but even on the tunnel level, or about 1,200 feet below the top of the lode, the sulphide ore is cut by veinlets of anglesite and thin stringers of manganese and iron oxides, which are at some places so closely spaced as to give the ore the appearance of being in a fairly advanced state of oxidation.

In a stope 60 feet above level 8, about 675 feet north of the shaft, bodies of ore of considerable size extend 15 or 20 feet from the main lode into the footwall. In this stope sulphide ore carrying quartz, galena, zinc blende, and pyrite is cut by many large stringers of ore composed of iron and manganese oxides, copper carbonate, lead carbonate, and sulphate. No barite was noted. The ore contains about 1 ounce of gold to the ton, associated with manganese oxides, and is much richer than the normal ore in the mine. On this level, south of the shaft, there is a body of oxidized ore, in places over 12 feet wide, which contains $5 in gold to the ton and some lead as sulphate and carbonate. A block of ground on the south end of this level is highly productive. In a small stope raised about 20 feet above the tunnel level, 400 feet north of the shaft, a crushed zone

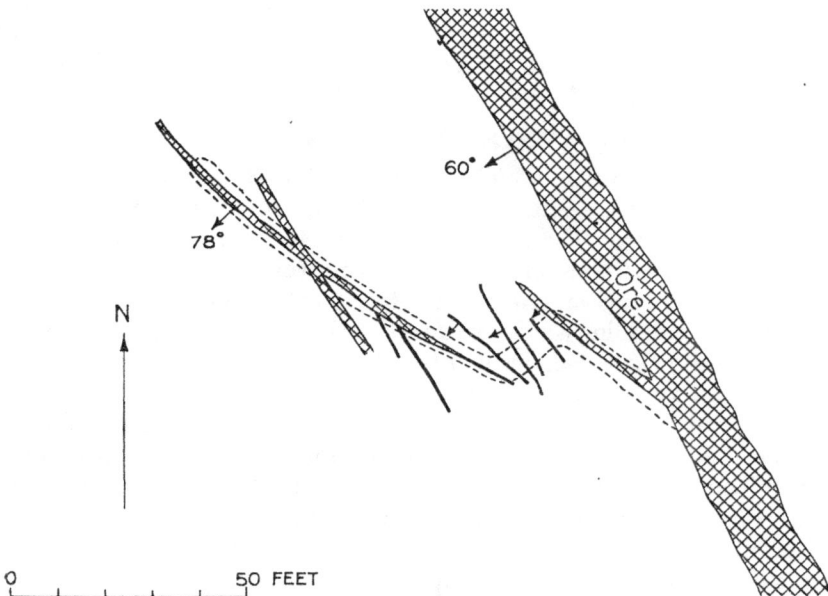

FIGURE 27.—Plan of part of level 7, Happy Thought mine, about 675 feet north of shaft.

some 8 feet wide is composed of rhyolite, quartz, chlorite, and sulphides. At this point, which is over 1,100 feet below the apex, the sulphide ore is cut by stringers of oxidized ore carrying limonite and wad. It is said to carry 8 per cent of lead and 10 ounces of silver and $2 in gold to the ton. Here also the ore containing the most gold is manganiferous.

On level 7, about 675 feet north of the shaft, is a zone of complex fracturing in the hanging wall of the fault. The details on this level are shown in figure 27. Some of the ore carries as much as 1 ounce of gold to the ton, but most of it is unworkable. On level 8, below this point, a vein in the hanging wall is followed 80 feet N. 58° W. and then 120 feet N. 43° W. Both walls are the Campbell Mountain rhyolite. The fissure is a sheeted zone 3 feet wide that dips 70° SW.

Veinlets cutting the ore carry iron and manganese oxides, with $3 to $8 in gold to the ton and 8 to 15 per cent of lead. Small stopes are raised in places 40 feet. This zone is presumably a fissure that was noted on levels 8 and 10, although on level 10 the fissure joins the vein 75 feet farther south or 600 feet north of the shaft. On level 10 this vein strikes northwest and dips 65° SW. Directly below this point, on level 11, a fissure in the hanging wall strikes N. 75° W. On level 12, 635 feet north of the shaft, a fissure in the Campbell Mountain rhyolite strikes N. 52° W. and dips about 72° SW. On levels 7 and 8, as already stated, a little ore has been stoped from this hanging-wall vein. It is not developed below level 9. Another small veinlet is exposed in a short crosscut on level 10, 425 feet north of the shaft, where a thin vein carrying about 2 inches of quartz and sphalerite strikes N. 56° W. and dips 60° SW.

WHITE STAR MINE.

The White Star mine, which is north of the Happy Thought and south of the Park Regent, extends along the Amethyst fault-fissure for about 750 feet. The surface workings are not extensive and in the summer of 1912 were inaccessible. The Humphreys tunnel passes through the White Star ground about 1,200 feet below the apex, and near the south end line a raise is run to level 11 of the Happy Thought, which has been extended through the White Star. The mine is under lease to the company that operates the Happy Thought, and in 1912 this company was stoping ore near the south end line on and above level 11. Plans were under way to extend the higher levels of the Happy Thought into the White Star ground. On the tunnel level the average strike of the fault in the White Star ground is N. 25° W., and it dips about 55° SW. The footwall is the Willow Creek rhyolite; the hanging wall, the Campbell Mountain rhyolite. The ore resembles that of the Happy Thought at similar depths, consisting in the main of quartz, chlorite, barite, galena, sphalerite, pyrite, and chalcopyrite, with some limonite and manganese oxide. The yield is mainly gold, with lead and zinc and a little silver. The ore is concentrated in the Humphreys mill, and the concentrates are shipped to smelters. The production of the mine is small.

PARK REGENT MINE.

The Park Regent mine is north of the White Star, and at the surface the shaft is about 2,000 feet N. 28° W. of the Happy Thought shaft. The mine was located in the early nineties, when a shaft was put down and some ore was stoped from the upper levels. In 1895 the owner, O. P. Poole, built a 10-stamp mill near the collar of the shaft, but according to report its operation was not satisfactory. Mr. Albert Collins estimates that the total production was not over $25,000 to 1912. In 1904 the mine passed into the hands of A. E.

Humphreys and associates, and the ownership is now incorporated as the Park Regent Mining Co.

The shaft, which connects with the Humphreys tunnel, is sunk to a depth of about 1,200 feet. The lower portion is an incline driven on the Amethyst fault. In the upper workings, which are not now accessible, two levels are reported to be driven northward from the shaft for 1,100 feet. In the lower workings level 11 of the Happy Thought was being extended in 1912 to the Park Regent shaft.

When visited, the mine was accessible only on the level of the Nelson or Humphreys tunnel, the north face of which is in the Park Regent mine, about 1,000 feet north of the shaft. The tunnel for this entire distance is driven on the Amethyst fault. Near the shaft the fault strikes about N. 20° W., but about 400 feet north of the shaft the fault curves slightly to the north, and the strike thus assumed continues to the face of the tunnel. The fault dips 53°–65° W. The footwall of the fault is the Willow Creek rhyolite and the hanging wall is the Campbell Mountain rhyolite. On the tunnel level the mineralization of the fault is comparable to that shown on this level in the more productive mines to the south. Near the shaft an ore shoot that is followed 160 feet along the strike carries, according to Mr. Albert Collins, 7 per cent of lead, 8 per cent of zinc, and $1.60 in gold and 2 ounces of silver to the ton. The vein is from 2 to 8 feet wide. At the face the vein carries 9 per cent of lead, 7 per cent of zinc, and $2.80 in gold and 2 ounces of silver to the ton. This ore is the green chloritic variety, like that of the Happy Thought mine, and contains very little barite. Some anglesite is developed in thin fractures in galena. A little amethystine quartz was noted at the face. Cerusite associated with manganese oxide was observed in the ore near the Park Regent shaft, but oxidation products on this level are comparatively rare. The ore in the higher levels is said to be richer than that on the tunnel level.

CAPTIVE INCA MINE.

The Captive Inca mine is on Deerhorn Creek about 6,000 feet north-northwest of the Park Regent mine. A shaft put down to a reported depth of 500 feet was inaccessible in 1912. Although the surface is covered near the shaft, it is thought that a fault of the Amethyst system passes near the collar. According to report, little or no ore was found. The faulting in this vicinity is discussed on page 90.

DOLGOOTH CLAIM.

The Dolgooth claim is on West Willow Creek near the mouth of Deerhorn Creek, on the road to the Equity mine. A shaft near the contact of latite and andesite is sunk 50 feet, and a crosscut to the east is said to strike a vein 17 feet from the shaft. The shaft was inaccessible when visited. On the dump some fragments of porphyritic latite, probably the tridymite latite, are impregnated with pyrite. According to the owner, Mr. A. R. Allen, some of the material carries several dollars in gold to the ton. About 600 feet northwest of this shaft another shaft, the Dolgooth No. 2, is sunk to a depth of 80 feet. This shaft also is now inaccessible, but to judge from the dump it was put down through shaly rhyolite tuff.

EQUITY MINE.

The Equity mine is on West Willow Creek about 1¾ miles a little west of north of the Captive Inca. By wagon road it is about 7 miles northwest of Creede. The country rock all belongs to the Alboroto group and includes the Willow Creek rhyolite, the Campbell Mountain rhyolite, and the Equity quartz latite. On the surface above

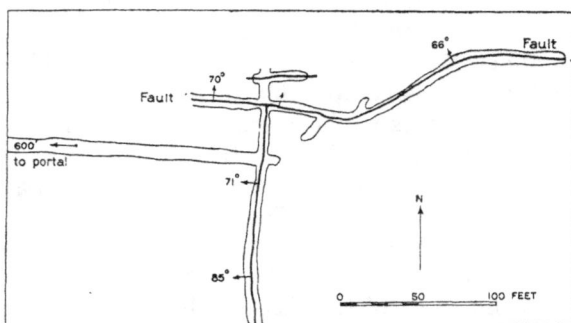

FIGURE 28.—Sketch of mine workings and principal fractures on tunnel level, Equity mine. Based on pace and compass survey.

the mine there is a great ledge of decomposed silicified rhyolite stained with iron oxides. The thin fracture seams in the rhyolite carry limonite, manganese oxide, and some barite. About 50 feet above the level of West Willow Creek a tunnel 725 feet long is driven S. 86° E. in the Campbell Mountain rhyolite. At about 715 feet from the portal drifts are driven to the north and south on a seam of sulphide ore 1 to 3 inches wide which carries considerable gold and silver. The drift to the north cuts the principal deposit, a fractured zone from 10 to 20 feet wide about 30 feet north of the tunnel. The principal fissure, as shown in figure 28, strikes east and dips 60°–70° N. The footwall is the Campbell Mountain rhyolite and the hanging wall the Willow Creek rhyolite. This fracture zone is a reverse fault with great throw. (See p. 92.) The rhyolite is highly silicified and impregnated with pyrite, zinc blende, and galena. It is followed eastward for 200 feet and locally carries considerable silver and gold. At the east face this fissure, shown in figure 28, carries a body of good silver ore about 2½ feet wide. No ore had been shipped from the Equity to 1912, but plans were under way to mine the high-grade ore that had been developed.

EXCHEQUER TUNNEL.

The Exchequer tunnel is about halfway between North Creede and the Commodore mine, in the rugged cliffs on the west side of West Willow Creek, about 600 feet above the bottom of the canyon. The country rock is rhyolite. On the surface 350 feet above the tunnel, near the contact between the Willow Creek and Campbell Mountain rhyolites, outcrops of shattered rhyolite show veinlets of quartz and barite carrying a little silver and gold. The tunnel is driven westward and northwestward about 1,100 feet in the Willow Creek rhyolite. Near the portal the flow lines strike northwest and dip about 50° SW. Toward the west the flow lines assume an attitude more nearly flat. Near its face the tunnel crosses the contact between the Willow Creek and Campbell Mountain rhyolites. This contact is followed east of north about 550 feet. It strikes, as shown in figure 29, about N. 30° E. and dips 25°–50° NW.

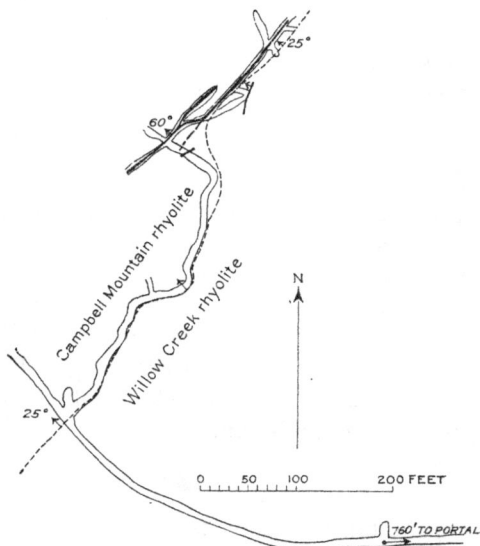

FIGURE 29.—Sketch of part of workings on tunnel level, Exchequer mine. Southeast of dotted line the country rock is the Willow Creek rhyolite; northwest of the line it is the Campbell Mountain rhyolite. The heavy lines are veins. Based on pace and compass survey.

There is much fracturing near the zone of contact but no clearly defined fault, the two formations being at some places tightly "frozen" together. That it is a normal flow contact is indicated also by the presence of numerous fragments of the Willow Creek rhyolite near the base of the Campbell Mountain rhyolite and a smaller number of such fragments at higher horizons in the Campbell Mountain rhyolite. These fragments were evidently broken from the Willow Creek rhyolite when the Campbell Mountain rhyolite flowed over it. The curving contact between the two formations suggests that the Campbell Mountain rhyolite flowed over a very irregular surface.

In the Campbell Mountain rhyolite there are three or four sheeted or fractured zones. All strike northeast and dip in general to the northwest. These zones consist of broken rhyolite, not highly altered by hydrothermal metamorphism, filled with barite gouge and some quartz. Here and there green copper stains impregnate the rock, and masses of quartz and galena carrying silver have been found. Much crosscutting and drifting has been done from the tunnel level, and some from a raise 35 feet above this level, but no workable body of ore had been encountered in 1912. (See fig. 29.)

The most persistent fissure strikes northeast and dips 60° to 90° NW. It is of variable width and has been followed about 240 feet along the strike. It contains much fragmental rhyolite and mud, with some quartz, barite, limonite, manganese oxide, copper carbonates, and native silver. According to Mr. Sidney Schroeder, one of the owners, it carries from 2 to 45 ounces of silver and averages $1.40 in gold to the ton. A little amethyst quartz is said to be present.

JO JO TUNNEL.

The Jo Jo tunnel is in North Creede, about 200 feet northwest of the end of the switch of the Denver & Rio Grande Western Railroad. It is run S. 83° W. 60 feet, then N. 80° W. for 90 feet. About 150 feet from the portal it is caved. The tunnel follows a slickensided surface which dips 65°–73° S. Near the portal the slickensided footwall shows prominent striae that plunge 85° W. in the plane of the vein, making an angle of 5° with the dip. The walls on both sides of the fissure are the Willow Creek rhyolite. Some quartz in the crushed rhyolite or gouge is said to contain a little silver, but no ore has been stoped.

MAMMOTH MINE.

The Mammoth mine is on Mammoth Mountain about 3,000 feet east of North Creede. Its elevation is 10,350 feet above the sea, and it is about 1,400 feet above East Willow Creek. No ore has been mined. Mr. William Barnett states that 10 tons of rich float yielding $900 was picked up in the early nineties on the slope between the Mammoth and the Nancy Hanks. A tunnel is driven about 325 feet southeast in a zone of crushed rhyolite and rhyolite flow breccia, and a winze 50 feet deep is sunk at the portal. Near the portal a large body of quartz was encountered, and this is followed in the workings along the tunnel for about 50 feet. The lode is a crushed zone which strikes about SE. and dips 55°–70° SW. From crosscuts in the walls it is estimated that the zone of crushing is about 25 feet wide. The footwall is the Willow Creek rhyolite and is typically fluidal. The hanging wall is brick-red rhyolite flow breccia of the Campbell Mountain rhyolite. The crushed zone, which clearly lies along a normal fault, extends to the face of the tunnel, a distance of 325 feet. It consists of fragments of rhyolite and rhyolite breccia, much gouge, and some fragments of quartz and barite. It is reported that this gouge carries about 10 ounces of silver to the ton. The minerals include white quartz, some amethystine quartz, barite, copper carbonate, silver chloride, and native silver. A green earthy mineral is probably chrysocolla. Some pieces of ore show projecting barite crystals surrounded by chalcedonic silica and quartz crystals. This vein, which is nearly on the strike of the Amethyst fault, is thought to be in the extension of the Amethyst fault system. (See p. 88.)

A second tunnel, the Nancy Hanks, 150 feet above the Mammoth tunnel, is driven 520 feet along the fault. This tunnel was not accessible in 1911 but is said to follow the vein that is exposed in the Mammoth tunnel.

The Eclat shaft, about 800 feet southeast of the portal of the Nancy Hanks tunnel, was not accessible in 1912. Some ore on the dump is said to carry 10 ounces of silver to the ton and a little gold.

OXIDE CLAIMS.

The Oxide claim group, which lies southeast of the Mammoth and the Nancy Hanks tunnels, is apparently located along the extension of the Amethyst fault system. The country is covered with débris, and outcrops are few, but without much doubt the fault continues through this group. In a shaft now inaccessible the lode is said to have been uncovered.

Where the rocks are exposed along the lode, extending from the Mammoth tunnel to the Oxide group, the northeast side or footwall is the Willow Creek rhyolite, and the southwest side the Campbell Mountain rhyolite.

PIPE DREAM CLAIM.

On the west slope of Dry Gulch the Mammoth fault can not be identified with certainty. A fissure that crosses Dry Gulch at an elevation of about 9,975 feet (aneroid) may be the extension of the Mammoth fault, but the rock on both sides is the Willow Creek rhyolite. A small tunnel on the Pipe Dream claim is driven along this fissure. Low assays are reported from the material along the fissure, but no marketable ore was found.

MUSTANG VEIN.

The Mustang vein is about 1,000 feet south of the portal of Bachelor tunnel 3. The portal of a tunnel driven on this vein has an elevation of about 9,925 feet. The vein strikes about N. 40° W. and dips 70°–85° SW. At the portal the country rock is the upper member of the Creede formation, but not more than 15 feet from the portal the tunnel encounters the red flow breccia of the Campbell Mountain rhyolite. This formation lies below the Creede formation. The contact shows no well-defined fissuring and is apparently an old erosion surface. The tunnel follows the vein northwestward for about 300 feet, then turns nearly at a right angle and runs south of west, crosscutting the hanging wall for a distance of about 220 feet. The vein is from 1 inch to 1 foot wide. It carries barite and limonite and a dark sulphide, which was not determined. It is said to contain silver. In the early history of the camp a carload of ore was shipped from this vein.

OVERHOLT LODE.

The Nelson tunnel, as already stated, is driven straight N. 68° W. about 2,117 feet. It was originally designed to cut several veins that crop out high above its level, but although it crosses several fractures, it encountered no workable ore. About 2,047 feet from the portal, or 70 feet from the breast, the tunnel crosses a thin vein that strikes N. 40° W. and is approximately vertical. A raise is run up about 300 feet on this vein, and short stopes and drifts are turned. On a level 97 feet above the Nelson adit the vein strikes N. 40° W. and dips 85° NE. The country rock is the Campbell Mountain rhyolite, and the lode is a fractured and sheeted zone about 2 feet wide. It is crossed by three small veins that strike N. 15° E. and dip 35°–70° SE. The main vein dips here about 85° NE. Movement is indicated by slickensides, gouge, and breccia, but as both walls are the Campbell Mountain rhyolite, the throw can not be determined.

In stopes above the 97-foot level the lode is a zone of fractured rhyolite cut by quartz and sulphides. Two 1-inch seams of finely ground sulphides and quartz parallel the walls. The ore runs 39 ounces of silver and 60 cents in gold to the ton. The minerals include white and amethystine quartz, barite, and an undetermined sulphide carrying silver and copper. The ore is partly oxidized and contains limonite and copper carbonate. A little chalcopyrite was noted.

At a point about 225 feet above the Nelson tunnel the direction of the dip changes. Below this point the lode dips northeast; above this point it dips southwest. On a short level about 300 feet above the adit, or 900 feet below the surface, the vein, which here is 5 feet wide, strikes N. 45° W. and dips 70°–85° SW. The fines and gouge, which are highly oxidized, carry 40 ounces of silver to the ton. In the summer of 1912 the mine was producing two carloads of ore a month.

RIO GRANDE NO. 2 LODE.

A small vein, known as the Rio Grande No. 2 lode, is cut in the Nelson adit about 205 feet north of the first turn of the adit, or about 1,600 feet from the portal. A raise said to be 175 feet long is driven on a distinct fissure that near the adit strikes N. 53° E. and dips 40° SE. Vein matter carrying barite and a little silver was found but no ore.

JACK POT LODE.

The Jack Pot lode is about 450 feet northwest of the Copper vein shaft and west of the Bachelor mine. The country rock is rhyolitic conglomerate of the upper member of the Creede formation. At the portal of the main tunnel the bedding is approximately horizontal. This tunnel is driven about 250 feet into the hill and has an average direction of N. 20° W. This vein is not far from the line of strike of a small vein near and parallel to the Mustang, but no connections have been discovered. The Jack Pot tunnel follows a zone of fracturing in the conglomerate and a short distance from the portal encounters a slickensided plane, which strikes about N. 10° W. This plane shows here and there a few inches of ore that contains quartz, limonite, and barite and is said to carry about 10 ounces of silver to the ton. Two winzes, one about 50 feet from the end of the tunnel and another at the end, are sunk on this baritic vein.

SOLOMON AND HOLY MOSES MINES.

The Solomon and Holy Moses mines are on the west side of East Willow Creek about 2 miles above Creede. The Solomon, which is about 1¼ miles north of North Creede, is separated from the Holy Moses by the Ridge mine. The Solomon and Holy Moses are the principal claims of the King Solomon Mining Co.; the Ridge is operated by a different company. The three mines are on the same lode, and they are connected by underground workings.

The deposit worked in the Holy Moses mine was discovered by N. C. Creede in August, 1889, and its discovery was largely instrumental in the rapid development of the Creede district. Some rich ore was taken out in the early nineties, and it is said that 1,065 tons had been mined up to December 31, 1892. This ore probably ran about $100 to the ton. Like many first discoveries this claim in point of production is dwarfed by comparison with later discoveries, near by. Although the Holy Moses shows some excellent ore near the surface, the deeper mining on this group has not been very profitable.

The Solomon was located by C. F. Nelson in 1890. Some very good ore was found near the surface, but the principal output of the mine came from the levels below the adit, which were turned from a deep shaft that is now under water. The Solomon and Holy Moses together have probably produced metals having a value of a little more than a million dollars, of which the greater part came from the Solomon mine.

The principal deposit is an anastomosing fissure vein that strikes about N. 7° W. and dips 55°–88° W. The underground workings on the Solomon group are extensive on three levels. The greatest developments are on the level of the Solomon adit, which is driven westward from the canyon of East Willow Creek at an elevation of about 9,300 feet. This tunnel encounters the vein about 420 feet from the portal. From the point of intersection a short drift is driven toward the south. Northward the vein is followed about 3,000 feet.

The Holy Moses No. 2 is a long crosscut adit, whose portal is about 3,700 feet north of the portal of the Solomon adit. It encounters the Holy Moses vein about 1,050 feet west of the portal. The elevation of the portal is 9,850 feet, or about 550 feet higher than the level of the Solomon adit. About 1,200 feet of drifts are run at this level. A short distance southwest of Holy Moses No. 2, about 500 feet higher and at an elevation of about 10,350 feet, the Holy Moses No. 1 tunnel is driven. At this level the vein is followed about 700 feet. The workings of this level are connected by a long upraise from the Holy Moses No. 2 adit. This upraise was not accessible in 1912.

The canyon of East Willow Creek is sunk in the Willow Creek rhyolite, and all the workings of the Solomon level and of the Holy Moses No. 2 are in this formation.

At tunnel 1 of the Holy Moses mine the footwall is the Willow Creek rhyolite and the hanging wall the Campbell Mountain rhyolite. Farther north the Campbell Mountain rhyolite forms both walls.

The Solomon vein crops out conspicuously on the steep slope of Campbell Mountain, about 300 feet above the portal of the lower adit. Here a zone of pronounced sheeting strikes a few degrees west of north and includes fissures that dip about 65° W. Some mining has been done in a small open cut. The ore is said to be a mixture of carbonates and sulphides, carrying silver and lead. South of this point, toward the creek, the vein is covered with talus, and it is not known to be exposed anywhere on the east side of East Willow Creek. The Homestake vein, east of East Willow Creek, dips west, like the Solomon vein, and if the Solomon vein continues toward the south with the same dip and strike it should pass near the Homestake vein. The nearest exposures on the two veins are almost three-quarters of a mile apart, however, and the country nearly everywhere between is covered with talus. North of the exposure above the portal of the Solomon adit the vein is encountered in several shallow pits and short tunnels but is not continuously exposed at the surface along the strike. Brecciation and slickensiding of the walls give evidence of movement, but here the faulting has involved but one formation. Sulphides appear along the outcrop at some places. At the Holy Moses shaft, above tunnel 1, the vein crops out at an elevation of about 10,500 feet, and its supposed apex continues at about that elevation to the Phoenix shaft, 2,850 feet north of the Holy Moses shaft.

On the Solomon adit, which is the lowest tunnel, the Solomon-Holy Moses vein system is developed for about 3,300 feet along the strike. This adit encounters the vein at about 420 feet from the portal (fig. 30). From this point the vein is followed on a short drift toward the south. About 75 feet north of this point the vein splits, inclosing a long and relatively narrow block of rhyolite. The east branch is termed the Ethel vein in the Solomon workings, and the west branch is termed the Solomon vein. The Ethel vein dips about 62° W. near the point of intersection; the Solomon vein is less steep. The Ethel vein is followed northward about 750 feet on the Solomon adit, and at the level of the Ridge adit, which is about 90 feet above the Solomon, this vein is developed for more than 1,000 feet along the strike. In the Ridge mine the east vein is termed the Ridge and the west vein the Mexico. The two branches diverge along their strike, and at a point 800 feet from the south intersection they are 300 feet apart. These branches probably join toward the north. The exact position of their intersection is not shown in the underground developments of the Solomon and Ridge mines, but, as estimated by projecting the two branches in the workings of the Ridge adit, it is probably 1,950 feet north of the south intersection. If the estimate is correct the block of rhyolite inclosed between the Solomon vein and the Ethel branch is 1,950 feet long and has a maximum width of about 300 feet. The Ridge and Solomon veins probably join near chute 9 on the level of the Solomon adit, but the intersection is not clearly exposed. At chute 9 the hanging wall of the Solomon dips 70° W. The raise on chute 9 is on a veinlet that dips 50° W. Up 50 feet above chute 10, grooves dip 70° S. in the hanging wall. Both veins dip as a rule 55°–75° W., but at some places are approximately vertical.

A deep inclined winze about 2,000 feet from the portal is sunk on the vein to a depth of about 420 feet. Four levels are turned at 100-foot intervals. The incline, which exploited the most highly productive portion of the mine, was under water when the mine was mapped. North of the incline the vein is followed on the Solomon adit level for about 1,600 feet. As shown in figure 30, the dip is nearly everywhere about 75° W., although locally it is somewhat steeper. At the north end of the level the vein strikes north and dips about 75° W. Above this point, at an elevation about 1,100 feet higher, the Holy Moses vein is developed from No. 1 adit level, where it dips 56°–66° W. Although the underground workings are not connected, it is thought that the Solomon and Holy Moses veins are on the same lode.

The ore of the Solomon and Ethel veins is composed of galena, zinc blende, pyrite, and a little chalcopyrite in a gangue of green chlorite, talc, and quartz. A little fluorite is present in some of the

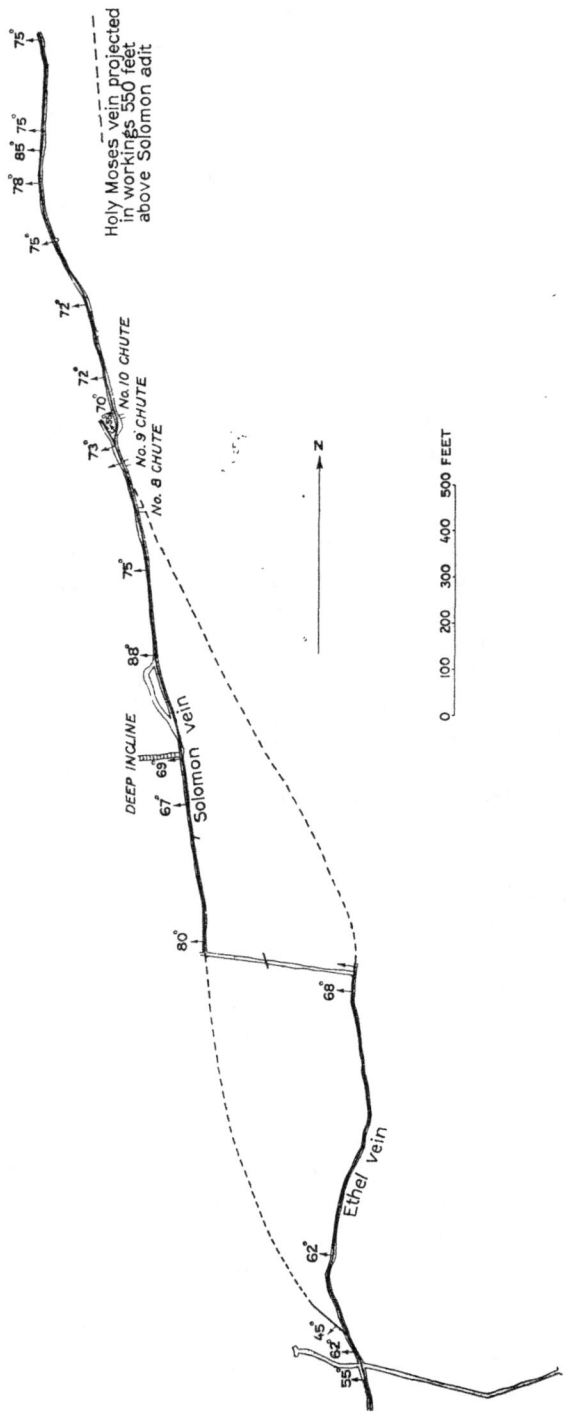

FIGURE 30.—Geologic map of Solomon adit, Solomon-Ridge-Holy Moses vein.

ore. It carries a small amount of gold, but its silver content is very low. In some of the sulphide ore gold and silver together are below $1 a ton, but the lead runs as high as 35 per cent. Barite is not abundant in the ore of the vein now exposed, although a little was found in the upper Holy Moses workings. Amethystine quartz was not noted, nor is quartz conspicuous in any of the ore. Extensive crushing has taken place since the ore was formed; indeed, both the Solomon and Ethel veins at most places are zones of green chloritic clay with abundant crystals of galena and zinc blende mixed with considerable crushed country rock. At some places the sulphides are powdered and mixed with a green mud resembling putty. There is probably considerable microscopic silica, but very little is visible in much of the ore.

Where exposed in the Solomon workings the vein is a great crushed zone, and a slickensided fissure runs nearly the whole distance through which the vein is developed. This fissure is in general on the hanging wall of the vein. About 1,150 feet north of the Solomon shaft or incline, in a raise 70 feet above the tunnel level, 9 feet of concentrating ore is exposed. This ore consists of zinc blende, galena, some pyrite, and a little limonite and manganese dioxide. The gangue is crushed rhyolite and green chloritic gouge,

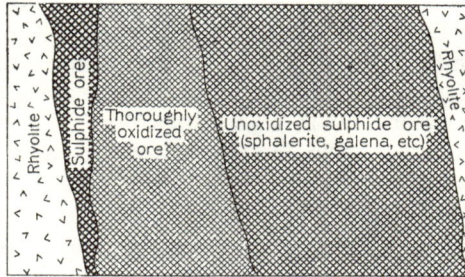

FIGURE 31.—Oxidized ore in sulphide ore from Solomon-Holy Moses vein 40 feet above Solomon adit, 3,000 feet from portal.

much like that in parts of the Amethyst vein. The ore carries 30 per cent of zinc and lead, with 1 to 3 ounces of silver and $1 to $2 in gold to the ton.

About 80 feet above the tunnel and 1,000 feet below the surface the sulphides change abruptly to oxides. The ore there consists of cerusite, limonite, manganese dioxide, a little galena and zinc blende, and considerable pyrite. This ore carries about $2.50 to the ton in gold. Two feet above the partly oxidized ore unaltered sulphides were noted. The occurrence of oxides along fissures and fractures far below the zone of complete oxidation is conspicuous, as it is also in the Amethyst vein.

At a point 3,000 feet from the portal and only 40 feet above the Solomon adit thoroughly oxidized ore is found in sulphide ore, as shown in figure 31.

RIDGE MINE.

The Ridge mine is on East Willow Creek between the Solomon and Holy Moses mines. The claim was located in 1890 by N. C. Creede and sold soon afterward to Senator Thomas Bowen. The principal adit is about a quarter of a mile north of the portal of the lower Solomon adit and is 90 feet higher. A tunnel is driven northwest 500 feet to the Ridge or Ethel vein, which is followed on the adit level a distance of 900 feet along the strike. From the drift on the Ridge vein a crosscut is driven to the Mexico or Solomon vein, and another crosscut about 600 feet long is driven southwest from a point near the north end of the drift on the Ridge vein. About 500 feet in from the portal of the lower tunnel of the Ridge a blind shaft is driven southwestward on the vein on an incline of 64°. Levels are turned at 90, 200, 250, 350, 450, and 500 feet. The workings in these levels below the adit aggregate about 2,500 feet, chiefly drifts on the vein. Below the Solomon adit the mine was submerged in 1912. A short tunnel is driven to the Ethel vein at a higher level, and stopes are raised on drifts that are run from the point of inter-section. This tunnel was not accessible in 1912. The mine is now worked only in a small way in levels not more than 200 feet above the adit. The ore in 1912 was treated in a small concentrating mill near the portal of the adit. The total production of the mine is said to be between $500,000 and $700,000.

The country rock is the Willow Creek rhyolite. The deposits are fissure veins of the replacement type. The Solomon and Ethel veins join in the Solomon mine about 2,700 feet from the portal of the lowest Solomon tunnel. They should join in the Ridge ground about 200 feet north of the north end of the drift on the Ridge vein, unless one or the other of these veins changes its present course to the north of this point. From this it is apparent that the Solomon-Mexico and Ethel-Ridge veins inclose a great horselike body of rhyolite that is probably about 1,850 feet long and over 300 feet wide at its widest place on the Ridge adit level. Both veins dip as a rule 55°–75° W., but at some places both are vertical. The Solomon vein is approximately vertical in a stope about 40 feet above the adit, between chutes 12 and 13, and 3,100 feet in from the portal of the Solomon adit. The Ridge vein is approxi-mately vertical 270 feet north of the shaft and 150 feet above the adit level.

The veins range in width from 1 or 2 feet to 15 feet. The ore of the veins is closely similar. It is composed of green chlorite and quartz, with galena, zinc blende, pyrite, and a little chalcopyrite. It carries a little gold and very little silver. The higher-grade ore contains nearly 35 per cent of lead and zinc sulphides. Barite is not abundant in the ore of the vein now exposed, and quartz visible to the unaided eye is almost lacking. Crushing has taken place since the ore was formed; indeed, both of the veins at several places observed are little more than zones of green gouge with crystals of galena and zinc blende, mixed with considerable crushed country rock. At several places the ore resembles a moist green putty containing powder of the sulphides.

On the Ridge vein about 250 feet north of the blind shaft that is driven from the lower tunnel level, 150 feet up in the stopes, the vein is from 10 to 15 feet wide. It dips 72° W. and has a smooth slickensided surface near the footwall. This surface is polished by movement and shows grooves inclined in the plane of the vein, making an angle of about 15° to the south of the line of steepest dip. The vein here consists of white and green mud, carrying many fragments of rhyolite not highly altered. Here and there streaks of nearly pure zinc blende and galena are included in the crushed mass, and crystals and lumps of powdered sulphides impregnate the mass. Locally, seams of limonite cross the ore, and some portions of the vein are oxidized. A mass of oxidized ore was found at the north end of the vein, 500 feet north of the blind shaft. Besides the two principal veins several fissures, or narrow crushed zones, were crossed by the adit between the portal and the blind shaft. One of these zones about 150 feet from the portal carries some lead and zinc sulphides but has been developed only a few feet north of the adit. A vein that crosses the adit about 50 feet east of the Ridge vein has been stoped for a short distance. Several minor fissures are exposed in the workings west of the Ridge or Ethel vein. Nearly all of them dip west at high angles. Except the Mexico or Solomon vein, none of them where exposed carry workable ore.

·PHOENIX MINE.

The Phoenix mine is on the east slope of Campbell Mountain, at an elevation of about 10,500 feet. A shaft 2,850 feet north of the Holy Moses No. 2 tunnel is sunk to a depth reported to be 120 feet. It is inclined 60° W. and is said to be driven on the vein. It has not been worked for many years. On the dump is crushed rhyolite containing galena, zinc blende, and pyrite. The Phoenix is near the strike of the Holy Moses vein and is probably its extension. On the broad flat south of the Phoenix shaft the Campbell Mountain rhyolite forms both walls of the Solomon-Holy Moses fault.

OUTLET TUNNEL AND ADA GROUP OF CLAIMS.

The Ada group of claims is on East Willow Creek, about 2 miles north of North Creede, east of the Solomon vein. The Outlet tunnel, which pierces Campbell Mountain at an elevation of 9,600 feet, is driven northwest for 1,172 feet and was designed to prospect two small veins that crop out on the ridge above. The country rock belongs to the Outlet Tunnel quartz latite.

About 300 feet from the portal the tunnel crosses a vein that strikes N. 40° W. and is nearly vertical. This vein carries 1 inch of rich ore composed of galena and zinc blende and said to contain 70 per cent of lead and 21 ounces of silver to the ton. At 440 feet from the portal there is a second vein that strikes west of north and carries a little silver. At a point about 950 feet from the portal the rhyolite is shattered and contains blowholes in which lead and zinc minerals occur. About 1,000 feet from the portal the tunnel encounters a third vein, which strikes nearly due north and dips 65°–85° W. This vein is followed along the strike for 260 feet. At one place 75 feet north of the point where it is crossed by the tunnel it is 4 feet wide and is said to carry 10 per cent of lead and about an equal amount of zinc. The sulphides, galena and zinc blende, are inclosed in green clay that fills spaces around fragments of the country rock. North of this point the vein is thin, at some places not over an inch wide. Southward from the main tunnel the vein is followed for 250 feet. In this portion it strikes S. 15° W. and dips steeply west. Some minor fractures are encountered in the workings west of this vein. About 1,057 feet from the portal of the tunnel a thin fracture dipping steeply west carries a seam of galena about half an inch wide.

The two small veins exposed on the cliff above are supposed to be cut in the tunnel. They are about 50 feet apart on the surface, and one is exposed at Holy Moses No. 2 ore bin. It is a very thin iron-stained fissure that strikes N. 20° W. and dips 54° W.

CARBONATE VEIN.

The Carbonate vein crops out high on the cliff northwest of the portal of the Outlet tunnel, about 1,600 feet north of the portal of Holy Moses tunnel No. 2, and is included in the Ada group of claims. The country rock is the Willow Creek rhyolite, which forms both footwall and hanging wall in the 100-foot tunnel that is run along the vein. Where the vein is exposed on the steep cliff of Campbell Mountain it strikes N. 17° W. and dips steeply west. It is a sheeted zone from 1 inch to 2 feet wide and carries small masses of quartz, jasper, lead carbonate, limonite, and hematite.

MOLLIE S. MINE.

The Mollie S. mine is high on the west slope of Mammoth Mountain, about 1 mile northeast of North Creede. The workings consist of four tunnels that have a difference in elevation of about 500 feet and aggregate about 1,300 feet in length. A tramway is built to deliver ore to a station on the wagon road on East Willow Creek from the dispatching station, 1,100 feet higher. According to Mr. Rufus Light, the owner, the mine has produced nearly $50,000.

The country rock is chocolate-colored rhyolite containing small white feldspar phenocrysts, and the flow lines at most places are approximately horizontal. It belongs to the Willow Creek rhyolite.

The lode is a fractured zone that strikes northwest and dips at high angles southwest. Slickensides and vein breccias are not highly developed. No faulting is determinable, for the same flow forms both walls. The ore consists of rhyolite shattered and seamed with native silver, cerargyrite, argentite, lead carbonate, chrysocolla, copper carbonate, chrysoprase, and a little galena. The chrysoprase is a beautiful green variety. Barite and amethystine quartz were not noted. Some of the rich ore consists of quartz highly impregnated with native silver. On some seams of rhyolite cerargyrite one-eighth of an inch thick plasters the rock fragments over a surface of several inches. Most of the ore that has been shipped consisted of fines from the material that was sorted over grizzlies. The rich silver minerals plastered on the fragments of country rock were knocked off in falling and passed through the screens. The barren or low-grade country rock was sent to the waste dumps. A noteworthy feature of the deposit is the nearly fresh condition of the rhyolite walls of the ore-bearing veinlets.

The chief ore body, which is explored from level 2, is developed for 125 feet along the strike. It extends downward 50 feet below this level and upward to level 1, about 75 feet vertically above. At one place a stope in this ore body is about 18 feet wide, and for a considerable portion of its length its width averages about 10 feet. The rhyolite is strongly sheeted, and the prominent planes of fracture trend north of west, with the lode. Tunnel 3, which is 125 feet lower than level 2 is driven north of east about 200 feet. It encounters a small mass of ore below the stope of level 2.

Tunnel 5 is about 320 feet below tunnel 3. This tunnel, which is driven eastward for 380 feet, encounters a thin vein about 230 feet from the portal. This vein, which is followed southward, strikes S. 10° E. and dips 60° W. It carries here and there a few inches of quartz.

EUNICE MINE.

The Eunice mine is directly south of the Mollie S. mine, and its principal tunnel is about 125 feet south of the portal of tunnel 2 of the Mollie S. The country rock is purple rhyolite, and here, as in the Mollie S. mine, the ore occurs as green silver-bearing copper minerals filling small fractures in the shattered rhyolite. In the cliffs above the portal of the tunnel the rhyolite is stained with green silver-bearing minerals, and it is said that several carloads of ore have been shipped, mainly from the surface.   The Eunice tunnel is driven eastward about 120 feet, then northwestward for 75 feet.   It intersects the same zone of fracturing that is developed in the Mollie S. mine. A raise is carried from the tunnel level about 35 feet to a small stope. The direction of the lode is not clearly shown in the tunnel, but the principal fractures strike about N. 55° W. and dip about 75° SW.   A small vein is exposed at several places on the surface about 100 feet southwest of the Mollie S.   This vein has not been stoped underground but is reported to have supplied some ore from surface workings.

DORA BELLE TUNNELS.

The Dora Belle prospects are about 700 feet south of the Mollie S. mine.   Two tunnels driven at about the same elevation and 55 feet apart are run 250 and 275 feet respectively into the rugged cliffs of Mammoth Mountain.   These tunnels are designed to cut either the Homestake or a parallel vein that crops out on the cliffs about 400 feet above the portals.   About 50 feet from the portal the north tunnel cuts a streak of black gouge that strikes east and dips 55° S. The mud from this streak is highly manganiferous, and according to the owner, Mr. G. T. Franks, it carries about 9 per cent of lead.   The south tunnel encounters near the breast a fractured zone that strikes N. 80° W. and dips 85° S.   The light-colored gougelike material in the fractures is said to carry a little lead and silver.

HOMESTAKE CLAIM.

The Homestake claim is about 1,600 feet northeast of the Mammoth mine, at an elevation of 10,500 feet.   A tunnel is driven southeastward for about 90 feet along a fractured zone in rhyolite.   For the greater part of this distance, extending to the breast, the tunnel follows a small vein carrying barite stained with limonite.   Several small fractures strike about N. 25° E. and dip 75°–80° W.   The Mammoth tunnel, which is driven S. 77° E. for about 2,280 feet toward this vein, should intersect it near the breast, but no orebearing fissure was identified in it.   A shaft, now inaccessible but said to have encountered ore, was sunk on Mammoth Mountain 600 feet north of the Homestake tunnel at an elevation about 200 feet higher.   In the steep cliffs of East Willow Creek canyon a pit is dug on a fractured zone, which strikes approximately with this vein. In the bottom of the pit there is about 6 inches of copper-stained baritic ore, said to carry silver.

MAMMOTH TUNNEL.

The Mammoth tunnel is on East Willow Creek about 2,500 feet above its junction with West Willow Creek.   It is driven S. 77° E. for about 2,280 feet in the Willow Creek rhyolite.   About 235 feet from its portal it encounters a small slickensided fissure that strikes N. 20° E. and dips 70° NW. About 860 feet from the portal is a second fissure that strikes N. 15° E. and about 925 feet from the portal another fissure that strikes N. 25° E. and dips 75° W. Between this point and the breast several other small barren fissures were noted. The tunnel, it is said, was first designed to cut the Mammoth vein, but the breast is probably farther from the vein than the portal. The Homestake lode, which crops out on Mammoth Mountain about 1,700 feet above the level of the tunnel, if its westward dip persists should be encountered in this tunnel, but it has not been identified.

RAMEY TUNNEL.

The Ramey tunnel, 1,000 feet north of the Mammoth tunnel, is driven S. 83° E. for 1,160 feet in the banded Willow Creek rhyolite. It crosses several small slips that strike about N. 10° E. and dip steeply west. No ore was seen.

MONTE CARLO MINE.

The Monte Carlo mine is on Campbell Mountain between the forks of East and West Willow creeks, about three-fifths of a mile north of North Creede. Its elevation is approximately 10,500 feet. A tramway is built from a tramhouse at the portal, at the elevation of the main level, to a station just north of the receiving station of the Mollie S. mine. This tramway is one of the steepest towerless trams in the West, its inclination being about 45° and its vertical fall about 1,350 feet.

On the surface several shallow shafts are sunk, and an open cut of moderate size is said to have produced ore. A crosscut tunnel driven south for 250 feet intersects the vein, which strikes about west and dips steeply south. From the tunnel a drift about 170 feet long is run on the vein. The drift is driven about 90 feet eastward from the point where the crosscut tunnel intersects the vein to the upper tramhouse on the steep cliff that forms the east slope of Campbell Mountain.

The deposit is a fractured and locally sheeted zone in streaky gray Willow Creek rhyolite. Copper carbonates, chrysocolla, and chrysoprase occur in the fractures, and this ore is said to carry silver. Although much money has been spent in equipping the mine with wagon road and tramway, it has produced but little ore. A striking feature of the deposit is the slight alteration of the wall rock near the ore.

It is thought by some that the Mollie S. and Monte Carlo deposits are on the same fractured zone. Although the country is the same formation and the minerals and ore are closely similar, there is little ground for the belief that they are on the same lode.

CONEJOS NO. 2 CLAIM.

The Conejos No. 2 claim is on the west slope of Windy Gulch about 100 feet below the wagon road from Creede to Bachelor, at an altitude of about 10,000 feet. The country rock is the Campbell Mountain rhyolite. A tunnel is driven N. 15° W. for 70 feet. About 40 feet from the portal it encounters a vein that strikes N. 7° E. and dips 78° W. The vein, which is 2 or 3 feet wide, is composed of barite and quartz and is said to carry 4 or 5 ounces of silver to the ton. North of this tunnel, at an elevation somewhat above it, a second tunnel was driven northwestward but did not encounter a vein.

CLEAVELAND TUNNEL.

The Cleaveland tunnel is near the bottom of Windy Gulch at an elevation of about 9,750 feet. The country rock is rhyolite of the Campbell Mountain formation. The tunnel is driven about northeast a distance of 510 feet. For a part of its length it follows a fissured zone that contains masses of barite here and there, with crushed silicified country rock.

TAHRSHATHEA CLAIM.

The Tahrshathea claim is on the north side of Windy Gulch about half a mile west of Willow Creek, at an elevation of about 9,350 feet. The country rock is Willow Creek rhyolite. From a vertical shaft 80 feet deep a 90-foot crosscut is driven west of north toward a ledge of rhyolite that crops out as a rugged cliff. The crosscut near the bottom of the shaft is run in mud containing abundant fragments of igneous rock. It encounters a contact of the banded rhyolite and loose material and follows it a short distance. According to the owner, Mr. A. C. Dean, no ore has been discovered.

BULLDOG, KANSAS CITY STAR, AND NORTH STAR CLAIMS.

The Bulldog, Kansas City Star, and North Star claims are on the east and north slopes of Bulldog Mountain, probably on the same fault. (See p. 93.) On the Bulldog claim a tunnel is driven 170 feet northwest to a slickensided fault fissure in rhyolite, which strikes N. 7° W. and dips E. 50°. On the hill above, on the Kansas City Star claim, a shaft was sunk 70 feet. According to Mr. W. H. Howell, one of the owners, samples running 87 ounces of silver to the ton were obtained from this shaft. It is now inaccessible.

On the North Star claim, on the north slope of Bulldog Mountain, a tunnel is run 275 feet along the fault, which strikes about S. 25° E. Crushed gouge carrying a little silicified material occurs along the fissure and is said to carry a few ounces of silver to the ton and some gold.

BETHEL CLAIM.

The Bethel claim is on the south slope of Bulldog Mountain, about 1¼ miles west-northwest of Creede. The country rock is the Campbell Mountain rhyolite, cut by dikes of intrusive rhyolite. On this claim the Oxford tunnel is driven 600 feet westward to the Bethel lode, which it intersects at 500 feet from the portal. On the top of a low spur of Bulldog Mountain, at an elevation of 10,000 feet, a shaft sunk 65 feet deep was not accessible in 1911.

The lode strikes N. 33° W. and dips 64° W. It is an oxidized zone of crushed rhyolite about 4 feet wide, and, according to the owner, Mr. F. G. Blake, it carries in places ore that contains 5.5 per cent of lead and 0.8 ounce of gold and 2 ounces of silver to the ton. About 8 feet southwest of the lode a dike of rhyolite porphyry cuts the rhyolite. The dike is 50 feet wide, strikes northwest, and dips steeply southwest. Its contact with the rhyolite is "frozen" and shows no evidence of shattering after intrusion. The face of the tunnel is in rhyolite.

LITTLE GOLD DUST CLAIM.

The Little Gold Dust claim joins the Bethel on the northwest, and the shaft, which is not now accessible, is about 350 feet northwest of the Bethel shaft. The dump shows rhyolite breccia and altered white porphyry. According to the owner, Mr. J. I. Howard, a small mass of manganiferous ore found in the shaft 34 feet below the surface was rich in gold.

ALPHA AND CORSAIR MINES.

The Alpha and Corsair mines are at Sunnyside, a small camp about 2 miles west-southwest of Creede. Sunnyside is nearer to the Rio Grande than Creede and therefore was more readily accessible to the pioneers who sought the San Juan Mountains by way of the Rio Grande Valley. It is currently reported that locations were made at Sunnyside as early as the seventies, 15 years or more before the discovery of the lodes at Creede, but if discoveries were made at so early a date the locations were allowed to lapse. It is stated on good authority, however, that the Alpha was located as early as 1883 by J. C. MacKenzie, who sold the claim soon afterward to Richard and J. N. H. Irwin. The Irwin Bros. located the Diamond I and Hidden Treasure claims, which are near the Alpha, and were engaged in prospecting and exploiting them until 1890. After the discovery of rich lodes at Creede the deposits at Sunnyside were worked with greater energy.

The Alpha and the Corsair were exploited by different companies, but they are on the same lode and have similar geologic features. The lode was exploited through three tunnels. The lowest one has an elevation of about 8,900 feet.   Entering the Alpha mine about 650 feet northeast of this tunnel and about 40 feet higher on the hill is a second long adit which is driven to the Corsair mine.   A third adit is about 125 feet above the lowest tunnel on the Alpha ground. About 825 feet from the portal of the Corsair tunnel a winze equipped with hoist and steam power is sunk to a depth reported to be 155 feet on an incline of about 52°.   In 1912 this winze was submerged 40 feet below the level, and the lowest tunnel of the Alpha was inaccessible.

The value of the total output of the two mines is estimated at $600,000, practically all in silver.   The greater part of this output was taken from the Corsair.   No ore was being mined in 1912, but the dump at the portal of the Corsair tunnel was being sorted by lessees and some very good ore was being found.

The country rocks are the Willow Creek rhyolite, the Campbell Mountain rhyolite, the bedded tuff of the Creede formation, and the quartz latite porphyry, which intrudes the rhyolites.   On the surface the vein is exposed near the point of a ridge between Miners and Rat creeks.   At this place it strikes a few degrees west of north.   It dips about 45°–60° NE. and follows a broad dike of porphyry.   At most places the porphyry dike is in the hanging wall of the vein, and here and there the vein is in the contact between the dike and the rhyolite, but at many places both walls are rhyolite.   The vein follows near the crest of the ridge for about 800 feet and, then, striking N. 30° W., passes on the northwest side of the ridge.   The fault fissure that it occupies may be followed almost continuously for about 6,000 feet to a point near the Kreutzer Sonata mine, and it is exposed in many shafts, tunnels, and pits.   Nearly everywhere on the surface the footwall is rhyolite, and at many places the hanging wall is porphyry. At the Yellow Jacket shaft, on the west slope of the south end of MacKenzie Mountain, no porphyry is exposed, and in a cut 400 feet northwest of this point both hanging wall and footwall are rhyolite.

The lode is a sheeted fault zone, at many places in or immediately below the contact of rhyolite and porphyry.   This contact nearly everywhere shows evidence of movement in slickensiding, grooving, and brecciation.   The zone of brecciation is at some places 20 or 30 feet wide and is made up essentially of gouge and fragments of crushed rhyolite and porphyry.   The porphyry crumbles very easily and weathers more rapidly than the rhyolite, and the fragments of rhyolite predominate in the fault zone.   The maximum sheeting is in the rhyolite, and at some places the contact between rhyolite and porphyry is tightly "frozen."   Such relations are shown at several places, one of them in the intermediate tunnel (adit No. 2), in the short crosscut in the hanging wall 370 feet northwest of the deep winze sunk below the tunnel.

The vein is essentially a replacement vein. It contains very little material showing crustified binding, ribbon structure, or other textures characteristic of deposits that are assumed to have filled open spaces. Some pieces of ore showing banded crusts were found on the dump, however, and these are probably fragments of material that filled open spaces. The ore now exposed in the mine, all of which is above the Corsair tunnel, is highly oxidized, and the waters now issuing from the lower Alpha tunnel deposit much limonite. Quartz, limonite, and green and blue sulphates, probably copperas and chalcanthite, are the most abundant minerals. Barite is present but is much less abundant than in the Amethyst vein. A little amethystine quartz was noted on the dump, but most of the quartz is massive white or gray and some is a chertlike variety, probably formed by replacement of rhyolite. Silver chloride is probably the most abundant ore mineral. Much pyrite is scattered through the ore on the dump and it doubtless is still more abundant in the lower levels. It is said that stibnite and stephanite were found also in the lower levels. A report of the president, C. S. Thompson, to the stockholders of the Corsair Mining Co., dated May 18, 1903, and describing operations for the preceding year, states that two ore shoots extend downward from the tunnel level. When the inclined winze was completed a drift was run north to a point below the north shoot, and from this point a raise was driven. It was found that the north shoot, which is well defined above, pinches out 6 feet below the water level. The south shoot was found 76 feet from the winze and, according to this report, all ore carrying more than 40 ounces of silver to the ton was taken out. About 40 feet of this ore shoot averaged 21 tons to the foot and yielded about $430 a foot. Some assorted ore, according to the report, carried 1,130 ounces of silver to the ton. Many consignments ran above 40 ounces.

The vein is near the contact of rhyolite and porphyry but is mainly in rhyolite. A straight crosscut (tunnel 2) is driven 540 feet northwest in rhyolite. At about 430 feet from the portal the tunnel crosses a small dike of porphyry. The tunnel turns 540 feet from the portal at a contact of rhyolite and porphyry and runs west for 130 feet in porphyry to a point where it encounters a zone of complex fracturing. From this point it follows a fissure 150 feet north to a deep winze. A drift on the vein is run 900 feet northwest of this winze, but the further extent of the drift is not known, as the workings are caved. In the portion of the vein northwest of the winze several stopes are carried upward from the level. From the winze to a point 375 feet northwest of it the vein is in rhyolite. Several short crosscuts are run to the contact. Although the rock along the contact is in places somewhat shattered it is not known to carry workable ores. About 450 feet northwest of the winze a small body of shattered porphyry is exposed in the hanging wall of the vein.   In the workings northwest of this point no porphyry was noted.

In the upper tunnel of the Alpha, about 85 feet above the Corsair, the vein where first encountered is in rhyolite.   About 40 feet beyond this point it enters porphyry, and for about 350 feet the tunnel is driven along the contact.   A short distance above the tunnel level, near the boundary between the Corsair and Alpha, three veins have been developed.   They are nearly parallel and probably intersect at very small angles.   The positions of the three veins of the sheeted zone on a level a few feet above the Alpha adit No. 2 are indicated on figure 5 (p. 109).   All these veins have been stoped at this place, but none of the stopes extend for great distances along the strike. Few data are available regarding sulphide enrichment or distribution of metallic minerals in depth.   The following analyses were supplied by the American Smelting & Refining Co.

Partial analyses of ore from Corsair mine.

[Per cent except as otherwise indicated.]

Date.	Tons.	Ag (ounces to the ton).	SiO₂.	Fe,Mg.	Al₂O₃.	Zn.	CaO.	S.
September, 1903	80	27.9	65.6	4	14.3	1.3	1	3.5
January, 1904	489	26.7	72.8	3		.3		3.6
August, 1904	127	27.1	66.8	4.6	15.2	1.7	1.5	4.5
January, 1903	294	62.4	74.8	3.1	12	2	1.3	3.3
August, 1903	329	45.2	65	4	15	1.7	1	3.6
September, 1894	29	50.6	75.2	3	6.9	.7	.7	.8

The ore represented by these analyses was mined during the later period of the history of the Corsair mine and was probably not so rich as ore taken from the mine in earlier years.   Inspection of these analyses and of some representing 38 other shipments shows that silica is in general a trifle higher in the richer ores.   In the richest ore aluminum is a little lower than in those of lower grade.   Iron and magnesium are a trifle higher in the low-grade ores.   Sulphur and iron do not increase in proportion to the increase of silver. These analyses appear to indicate that very little of the silver is in pyrite, and that the richer silver ore is in general highly siliceous. These ores were mined, in the main, below the water level.   The analyses do not show any clearly defined relation between the state of oxidation and the tenor.   No oxide enrichment is indicated by these meager data.

RENO PROSPECT.

The Reno shaft is about three-quarters of a mile northwest of Sunnyside, at an elevation of about 9,250 feet.   The shaft is sunk 55 feet on a fissure that strikes N. 28° W. and dips 60° NE.   No ore was noted in the shaft.   Both walls of the fissure are porphyry, but rhyolite is exposed in a ledge a few feet west of the fissure. The Alpha fault is probably a few feet west of the Reno shaft; and the fissure on which the shaft is sunk is a parallel fissure in the hanging wall of the fault.

About 100 feet below the collar of the shaft a tunnel is run 160 feet about N. 60° E. The portal is in rhyolite, but the tunnel encounters the contact between rhyolite and porphyry at 20 feet from the portal and continues in porphyry 140 feet. About 135 feet from the portal the tunnel crosses a fracture that strikes about northwest and dips 55° NE. A thin veinlet of silver ore fills the fracture here and there. A few paces northwest of this point and about 75 feet lower, a second tunnel is driven on a contact between rhyolite on the footwall side and porphyry on the hanging-wall side. The contact, which is obviously a fault, at this place dips 58° NE.

KREUTZER SONATA MINE.

The Kreutzer Sonata mine is on Miners Creek about 1¼ miles above Sunnyside. It was opened in 1892 and is said to have produced about $10,000, but it has not been worked for several years. Three tunnels are driven on the lode. The lowest tunnel (1) is near the level of Miners Creek, tunnel 2 is about 80 feet higher, and the highest tunnel is 125 feet above tunnel 2. These tunnels are 400, 480, and 200 feet long, respectively. The lowest tunnel was not accessible when the mine was visited in 1912.

The prevailing country rock is the Willow Creek rhyolite which is intruded by porphyry. In tunnel 2, near its portal a broad dike of porphyry cuts the rhyolite. This porphyry is similar to that exposed in the Corsair and Alpha mines. The contact of rhyolite and porphyry strikes northwest and is nearly vertical. It is slickensided and has clearly been subject to movement since the porphyry was intruded into the flow.

The deposit is a fissure vein in rhyolite. It does not crop out conspicuously. On the surface above tunnel 3 a veinlet in rhyolite, not more than 1 inch wide, is exposed. In tunnel 2 the lode which strikes about N. 60° E. and dips 60°–80° SE., is 2 or 3 feet wide, and the rock is considerably altered. This tunnel follows the lode for about 225 feet. A stope is raised a short distance above the level, and an incline and a winze are put down about 30 feet below the level directly under the stope. The ore contains zinc blende, galena, and pyrite and is said to have carried silver and gold. Considerable crushing has taken place since the vein was deposited, and the ore contains much gouge. Mr. M. J. Le Fevre states that picked ore ran over 1,000 ounces of silver to the ton. The minerals were galena, wire silver, and probably argentite.

DIAMOND KING CLAIM AND PARIS TUNNEL.

A few rods north of the Kreutzer Sonata mine, in the small gulch that enters Miners Creek at an elevation of about 9,025 feet, the Willow Creek rhyolite and porphyry are complexly fissured and faulted. Although iron stains indicate mineralization along the fissures, no ore body has been developed. A tunnel on the Diamond King claim is driven northwest about 100 feet on a steeply dipping fracture in purple rhyolite. Black earthy sulphides that occur here and there in the fractured zone are said to be rich in silver. Lower on the hill, about 50 feet above Miners Creek, the Paris tunnel is driven 600 feet N. 54° E. to intersect these fissures in depth. It crosses rhyolite and porphyry that are probably in faulted contact. Small veins striking N. 7° W. are crossed 225 feet and 300 feet from the portal. About 500 feet from the portal a crushed zone striking N. 30° W. is encountered. The rock is much decomposed, and according to one of the owners, Mr. M. J. Le Fevre, it carries gold and silver.

MONON MINE.

The Monon mine, owned by the Quintette Mining Co., is on the west slope of Monon Hill a few rods east of Sunnyside. This hill is composed chiefly of tuffs and limestones of the Creede formation. The details of the geologic structure are complex, for the limestones are probably spring deposits, partly recrystallized, and they were probably deposited in openings as well as on the surfaces of the tuffs between which they are now interbedded. The series has been folded and faulted since it was deposited. The lowest member exposed in the underground workings is a bed of rhyolitic tuff and breccia. Above this is a thin band of siliceous shaly material, probably in part a tuff and in part silicified calcareous material. A few feet of soft shaly tuff lies above this material and above the tuff is rhyolite breccia like that of the lowest member exposed. Owing to the nature of their origin not much confidence can be placed in the calcareous beds as horizon markers.

Near the top of the hill several shafts and short tunnels are run, in the main along contacts of limestones and tuffs. An adit (fig. 32) at an elevation of about 9,175 feet is driven S. 68° E. about 700 feet. At 200 feet from the portal a drift is driven northeast about 100 feet, then southeast 110 feet. The drift follows the contact of the lower rhyolite tuff and the overlying shaly material, which lie in the steeply plunging end of an anticline. Two raises follow the contact upward, and about 100 feet above the tunnel irregular workings follow a flat-lying bed of the silicified shaly tuff.

The ore occurs mainly in altered breccia below the shaly tuff and in silicified and calcareous material that lies at some places between the tuff and the breccia. The breccia is fissured and seamed with iron oxide, which is said to carry silver chloride and other silver minerals. In the main tunnel, about 350 feet from the portal, a 1-inch veinlet containing a dark sulphide dips 70° NE. According to one of the owners, Mr. Charles P. Eades, about 5 carloads of ore running from 25 to 71 ounces of silver to the ton have been shipped from the mine. The ore is said to carry no lead, zinc, or gold.

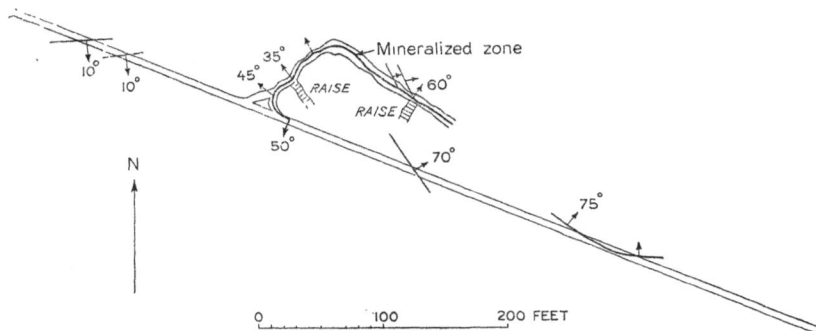

FIGURE 32.—Sketch of the adit level, Monon mine. Based on pace and compass survey.

SUNNYSIDE TUNNEL.

The Sunnyside tunnel is on Rat Creek about three-quarters of a mile north of Sunnyside. It was designed to cut some thin veins presumed to have supplied rich float that was found on MacKenzie Mountain. It is run 1,100 feet about N. 80° W. and crosses four thin fractures that strike N. 7°–28° W. and stand nearly vertical. According to Mr. D. Brennan, the owner, no ore has been encountered.

DELAWARE CLAIM.

The Delaware claim is on Rat Creek about 1,250 feet above the Sunnyside tunnel. It was located in 1894 by John C. MacKenzie, and, according to report, a small shipment of rich ore was made. An oxidized zone of fracturing is exposed 100 feet above Rat Creek. It strikes N. 30° W. and dips 65° E. A shaft 80 feet deep is sunk on the lode, and several pits have been dug along the strike of the lode. The dump at the shaft shows considerable highly iron-stained rhyolite and some pyrite and marcasite. In 1912 Charles Lee had a bond and lease on this claim and was driving a tunnel to cut the lode in depth.

www.ingramcontent.com/pod-product-compliance
Lightning Source LLC
Chambersburg PA
CBHW022055210326
41519CB00054B/454